CONSTRUCTION MANUAL:

CONCRETE & FORMWORK

By T. W. LOVE

Craftsman Book Company
6058 Corte Del Cedro, Carlsbad, CA 92008

ACKNOWLEDGEMENT

The author is indebted to the Portland
Cement Association for permission to
use copyrighted P.C.A. materials in the
preparation of this manual.

CONTENTS

CHAPTER 1

CONCRETE AS A BUILDING MATERIAL

Portland cement is universally considered to be the most important masonry material used in modern construction. Its numerous advantages make it one of the most economical, versatile, and universally used construction materials available. It is commonly used for buildings, bridges, sewers, culverts, foundations, footings, piers, abutments, retaining walls, and pavements. A concrete structure, either plain or reinforced, is almost unique among the many systems of modern construction. In its plastic state concrete can be readily handled and placed in forms and cast into any desired shape. Quality concrete work produces structures which are lasting, pleasing in appearance, and require comparatively little maintenance.

Limitations. Recognition of the limitations of concrete construction in the design phase will eliminate some of the structural weaknesses that detract from the appearance and serviceability of concrete structures. Some of the principal limitations and disadvantages are:

(1) *Low tensile strength.* Concrete members which are subjected to tensile stress must be reinforced with steel bars, high-strength steel wire, or mesh.

(2) *Drying shrinkage and moisture movements.* Concrete, like all construction materials, contracts and expands under various conditions of moisture and/or temperature. This normal movement should be anticipated and provided for in the design, placement, and curing. Otherwise, damaging cracks may result.

(3) *Permeability.* Even the best concrete is not entirely impervious to moisture. It contains soluble compounds which may be leached out to varying degrees by water. Impermeability is particularly important in reinforced concrete where reliance is placed on the concrete cover to prevent rusting of the steel, and where the structure is exposed to freezing and thawing.

CONCRETE COMPONENTS

Chemical Process. The essential ingredients of concrete are cement and water which react chemically in a process called hydration to form another material having useful strength. Hardening of concrete is not the result of the drying of the mix, as can be seen from the fact that fresh concrete placed under water will harden despite its completely submerged state. The mixture of cement and water is called cement paste, but such a mixture, in large quantities, is prohibitively expensive for practical construction purposes and undergoes excessive shrinkage upon hardening.

Aggregates. Inert filler materials in the form of sand, stone, and gravel are added to cement and water in prescribed amounts to increase the volume of the mixture. When concrete is properly mixed each particle of aggregate is completely surrounded by paste and all spaces between aggregate particles are completely filled. The paste is the cementing medium that binds the aggregate particles into a solid mass.

Grout. Grout is a mixture of portland cement, lime, fine aggregate, and water in such proportions that the mixture is fluid. Exact proportions and the maximum size of the aggregate are dictated by the intended purpose.

Mortar. Mortar is a mixture of portland cement, lime, fine aggregate, and water in such proportions that the mixture is plastic. Exact proportions and the maximum size of the aggregate are determined by the intended purpose.

PROPERTIES OF CONCRETE

A plastic concrete is a concrete mix that is readily molded, yet changes its shape slowly if the mold is immediately removed. The degree of plasticity influences the quality and character of the finished product. Control of the ingredients in the mix limits the variables to the proportions of the ingredients. Significant changes in the mix proportions are indicated by the slump. The desirable qualities of plastic concrete are:

Workability. This property indicates the relative ease or difficulty of placing and consolidating concrete in the form. The consistency of the mixture is measured by the slump test (Appendix A) and is maintained as necessary to obtain the required workability for the specific conditions and method of placement. A very stiff mix would have little slump and would be very difficult to place in heavily reinforced sections. It is a good mix to place in a slab where reinforcing is not used. A more fluid mix can be placed where reinforcing steel is present. Workability is controlled largely by the amounts of and proportion of fine to course aggregate used with a given quantity of paste.

Nonsegregation. A plastic concrete should be handled so that there will be a minimum of segregation and the mix will remain homogeneous. For example, to prevent segregation, plastic concrete should not be allowed to drop (free fall) more than 3 to 5 feet. Care must also be taken in handling to prevent bleeding.

Uniformity. For uniformity every batch should be accurately proportioned according to the specifications. Uniform quality of the hardened concrete is desirable from both economic and strength considerations.

HARDENED CONCRETE

Hardened concrete in finished form is the actual basis of any concrete design. The essential qualities which must be considered are:

Strength. Strength is the ability of the concrete to resist a load in compression, flexure, or shear. The principal influencing factor on strength is the ratio of water to cement. About 2½ gallons of water are required for hydration (chemical reaction with water) of a sack of cement. Additional water is used to thin the paste, thus allowing it to coat more particles. This increases the yield obtainable from each sack of cement, thereby producing a more economical mix. However, excessive water-cement ratios should be avoided because thin paste is weak and a reduction in strength of the hardened concrete occurs due to the dilution of the paste. The minimum and maximum amounts of water generally used for economical mixes range from 4 to 8 gallons per sack.

Durability. Durability in concrete is the ability of the hardened mass to resist the effects of the elements, such as the action of wind, frost, snow, and ice, the chemical reaction of soils, or the effects of salts and abrasion. Durability is affected by climate and exposure. As the water-cement ratio is increased, the durability will decrease correspondingly. Air-entrained cements produce concretes with improved durability.

Water-Tightness. Water-tightness is an essential requirement of concrete. Tests show that the water-tightness of the paste is dependent on the amount of mixing water that is used and the extent to which the chemical reactions between cement and water have progressed. Frequently specifications for water-tightness limit the amount of water used in concrete mixes to 6 gallons per sack of cement. Water-tightness of air-entrained concrete is superior to that of non-air-entrained concrete.

2

CHAPTER 2

CONCRETE MIX COMPONENTS

NATURAL CEMENT

Natural cement is produced by grinding and calcination of a natural cement rock which is a clayey limestone containing up to twenty-five percent of clayey material.

Characteristics of Natural Cement. Natural cement is normally yellow to brown in color. The tensile strength and compressive strength of natural cement mortars are low, varying from one-third to one-half of the strengths of normal portland cement. Natural cement is variable in quality and is little used today. In the United States, it represents no more than one percent of the production of all cements.

Uses of Natural Cement. Two types are available commercially: Type N natural cement and Type NA air-entraining natural cement. Natural cement is used in the preparation of masonry cements for use in mortar and in combination with portland cement for use in concrete mixtures.

PORTLAND CEMENT

Description. Portland cements are mixtures of selected raw materials which are finely ground, proportioned and calcined to the fusion temperature (approx. 2700°F.) to give the desired chemical composition. The clinker resulting from calcination is then finely pulverized. When combined with water, these cements undergo a chemical reaction and harden to form a stone-like mass. This reaction is called hydration and these cements are termed hydraulic cements.

Manufacture. Raw materials used in the manufacture of portland cements may include limestone, cement rock, oyster shells, coquina shells, marl, clay, shale, silica sand, and iron ore. These materials are pulverized and mixed so that the appropriate proportions of lime, silica, alumina, and iron components are present in the final mixture. To accomplish this, either a dry or wet process is used. In the dry process, grinding and blending are done with dry materials. In the wet process, the grinding and blending operations utilize a watery slurry. The prepared mixture is then fed into a rotary kiln which produces temperatures of 2600° to 3000°F. During this process, several reactions occur which result in the formation of portland cement clinker. The clinker is cooled and then pulverized with a small amount of gypsum added to regulate the setting time. The pulverized product is the finished portland cement. It is ground so fine that nearly all of it will pass through a sieve having 200 meshes to the lineal inch or 40,000 openings in a square inch. Each manufacturer of portland cement uses a trade or brand name under which the product is sold.

Common ASTM Portland Cements. Different types of portland cements are manufactured to meet certain physical and chemical requirements for specific puroses. The American Society for Testing and Materials (ASTM) provides for five types of portland cement in ASTM C150, "Standard Specifications for Portland Cement".

(1) *ASTM Type I.* This type is also called normal portland cement and is a general-purpose cement suitable for all uses when the special properties of the other types are not required. Type I portland cement is more generally available than are the other types of cement. It is used in pavement and sidewalk construction, reinforced concrete buildings and bridges, railways, tanks, reservoirs, sewers, culverts, water-pipes, masonry units, and soil cement mixtures. In general, it is used when concrete is not subject to special sulfate hazard or where the heat generated by the hydration of the cement will not cause an objectionable rise in temperature.

(2) *ASTM Type II.* This type is a modified cement used where precaution against moderate sulfate attack is important, as in drainage structures where the sulfate concentrations in the soil or ground water are higher than normal, but not unusually severe. Type II will usually generate less heat at a slower rate than Type I. It may be used in structures of considerable size where cement of moderate heat of hydration will tend to minimize temperature rise, as in large piers, heavy abutments, and heavy retaining walls and when the concrete is placed in warm weather.

(3) *ASTM Type III.* Type III is a high-early-strength cement which provides high strengths at an early period, usually a week or less. Concrete made with Type III cement has a 7-day strength comparable to the 28-day strength of concrete

3

made with Type I cement, and a 3-day strength comparable to the 7-day strength of concrete made with Type I cement. Type III cement has a higher heat of hydration and is more finely ground than Type I cement. It is used where it is desired to remove forms as soon as possible, to put the concrete in service as quickly as possible, and in cold weather construction to reduce the period of protection against low temperatures. Although richer mixtures of Type I may be used to gain high-early-strength, Type III may provide it more satisfactorily and/or more economically.

(4) *ASTM Type IV*. Type IV is a low heat cement for use where the rate and amount of heat generated must be minimized. The development of strength is also at a slower rate. It is intended for use only in large masses of concrete such as large gravity dams where temperature rise resulting from the heat generated during hardening is a critical factor.

(5) *ASTM Type V*. Type V is a sulfate-resistant cement used only in concrete exposed to severe sulfate action. It is used principally where soil or groundwater in contact with the concrete structure has a high sulfate content. It gains strength more slowly than Type I.

Other ASTM Portland Cements. The American Society for Testing Materials covers certain other types of portland cements in separate ASTM specifications. These types are:

(1) *Air-entraining portland cements.* Specifications for three types—Types IA, IIA, and IIIA—are given in AST MC175. They correspond in composition to Types I, II, and III, respectively in ASTM 150 with the addition of small quantities of air-entraining materials interground with the clinker during manufacture. These cements produce concrete with improved resistance to freeze-thaw action and to scaling caused by chemicals applied for snow and ice removal. Such concrete contains minute, well-distributed, and completely separated air bubbles.

(2) *White portland cement.* White portland cement conforms to the specifications of ASTM C150 and C175. The finished product is white instead of gray. It is used primarily for architectual purposes.

(3) *Portland blast-furnace slag cements.* The cements include two types conforming to the requirements of ASTM C595—Type IS and Type IS–A. To the latter, air-entraining additive has been added. In producing these cements, granu-

lated blast furnace slag is either interground with portland cement clinker or blended with portland cement. These cements can be used in general concrete construction.

(4) *Portland-pozzolan cements.* Portland-pozzolan cements include four types (P, IP, P–A, and IP–A, the latter two containing an air-entraining additive) as specified in ASTM C595. In these cements, pozzolan consisting of siliceous, or siliceous and aluminous material is blended with ground portland cement clinker. They are used principally for large hydraulic structures such as bridge piers and dams. The comparative strength of concrete made with portland-pozzolan cements may be lower than that made with normal cements.

(5) *Masonry cements.* Masonry cements are mixtures of portland cement, air-entraining additives, and supplemental materials selected for their ability to impart workability, plasticity, and water retention to masonry mortars. These cements conform to the requirements of ASTM C91.

Special Portland Cements. In addition to the above cements, there are special types of portland cement not covered by ASTM specifications.

Oil well cement. Oil well portland cement is made to harden properly at the high temperatures prevailing in very deep oil wells.

Waterproofed portland cement. Waterproofed portland cement is made by grinding water-repellent materials with the clinker from which it is made.

Plastic cements. Plastic cements are made by adding plasticizing agents to the clinker. It is commonly used for making mortar, plaster, and stucco.

Packaging and Shipping. Cement is shipped either in sacks which weigh 94 pounds and have a loose volume of 1 cubic foot or in bulk by railroad, truck, or barge. Cement requirements for large projects may be given in terms of barrels, a barrel being equivalent to 4 sacks or 376 pounds.

Storage. Portland cement that is kept dry retains its quality indefinitely. Cement which has been stored in contact with moisture sets more slowly and has less strength than dry cement. Sacked cement should be stored in a warehouse or shed as nearly airtight as possible. The sacks should be stored close together (to reduce the circulation of air) and away from outside walls. Bags to be stored for long periods should be covered with tarpaulins or other waterproof covering. If no shed is available, the sacks should be placed on raised wooden platforms. Waterproof coverings should be placed over the pile in such a way that rain cannot reach the cement or the platform. Occasionally, sacked cement in storage will develop what is commonly called "warehouse pack", a condition resulting from packing too tightly. The cement retains its quality under these conditions and the condition can usually be corrected by rolling the sacks on the floor. Cement should be freeflowing and free of lumps. If lumps exist that are hard to break up, the cement should be tested to determine its suitability. Hard lumps indicate partial hydration and will reduce the strength and durability of the concrete. Bulk cement is usually stored in weatherproof bins. Ordinarily, it does not remain in storage very long but it can be stored for a relatively long time without deterioration.

WATER IN THE MIX

The purpose of water in concrete mix is to combine with the cement in the hydration process, coat the aggregate, and permit the mix to be worked.

Mixing water should be free from organic materials, alkalies, acids, and oil. In general, water that is fit to drink is suitable for mixing with cement. However, water with excessive quantities of sulfates should be avoided even though it may be fit to drink. Otherwise, the result is a weak paste that may contribute to deterioration or failure of the concrete. Water of unknown quality may be used for making concrete if mortar cubes made with this water have 7- and 28-day strengths equal to at least 90 percent of companion specimens made with drinkable water. Tests should also be made to be sure that the setting time of the cement is not adversely affected by impurities in mixing water. Impurities in mixing water, when excessive, may affect not only setting time, concrete strength, and volume constancy, but may cause efflorescence or corrosion of reinforcement. In some cases, it may be necessary to increase the cement content of the concrete to compensate for the impurities. The effects of certain common impurities in mixing water on the quality of plain concrete are given below:

Alkali Carbonate and Bicarbonate. Carbonates and bicarbonates of sodium and potassium may either accelerate or retard the set of different cements. In large concentrations these salts can materially reduce concrete strength. Tests should be made when the sum of these dissolved salts exceeds 1,000 ppm (parts per million).

Sodium Chloride and Sodium Sulfate. A high dissolved solids content of a natural water is usually the result of a high content of sodium chloride or sodium sulfate. Concentrations of 20,000 ppm of sodium chloride and 10,000 ppm of sodium sulfate are generally tolerable.

Other Common Salts. Carbonates of calcium and magnesium are seldom found in sufficient concentration to affect the strength of concrete. Bicarbonates of calicum and magnesium may be present in concentrations up to 400 ppm without adverse effect. Magnesium sulfate and magnesium chloride can be present in concentrations up to 40,000 ppm without harmful effects on strength. Calcium chloride may be used to accelerate both hardening and strength gain.

Iron Salts. Natural groundwater usually contains only small quantities of iron. However, acid mine waters may contain large quantities of iron. Iron salts in concentrations up to 40,000 ppm can be tolerated.

Miscellaneous Inorganic Salts. Salts of manganese, tin, zinc, copper, and lead may cause a significant reduction in strength and cause large variations in setting time. Sodium iodate, sodium phosphate, sodium arsenate, and sodium borate can greatly retard the set and the strength development. Concentrations of these salts up to 500 ppm are acceptable in mixing water. Concentrations of sodium sulfide of even 100 ppm may be harmful.

Seawater. Seawater containing up to 35,000 ppm of salt is generally suitable for unreinforced concrete. Some strength reduction occurs but this may be allowed for by reducing the water-cement ratio. The use of seawater in reinforced concrete may increase the risk of corrosion of reinforcing steel; however, this risk is reduced if the reinforcement has sufficient cover and if the concrete is watertight and contains an adequate amount of entrained air. Seawater should not be used for making prestressed concrete in which the prestressing steel is in contact with the concrete.

Acid Waters. If possible, the acceptance of

acid mixing water should be based on the concentration (in parts per million) of acids in the water rather than the pH of the water. Hydrochloric, sulfuric, and other common inorganic acids in concentrations up to 10,000 ppm generally have no adverse effect on concrete strength.

Alkaline Waters. Concentrations of sodium hydroxide above .5 percent by weight of cement may reduce the concrete strength. Potassium hydroxide in concentrations up to 1.2 percent by weight of cement has little effect on the concrete strength developed by some cements but substantially reduces the strength of other cements. If in doubt, tests should be made.

Industrial Waste Waters. Industrial waste waters carrying less than 4,000 ppm of total solids generally produce a reduction in compressive strength of no more than 10 percent. Water that contains unusual solids, such as those from tanneries, paint factories, coke plants, chemical and galvanizing plants, and so on, should be tested.

Waters Carrying Sanitary Sewage. Sewage which has been diluted in a good disposal system generally has no significant effect on the concrete strength.

Sugar. Small amounts of sugar, .03 to .15 percent weight of cement, usually retard the setting of cement. When the amount is increased to about .20 percent weight of cement, the set is usually accelerated. Sugar in quantities about .25 percent by weight of cement may cause substantial reduction in strength. If the concentration of sugar in the mixing water exceeds 500 ppm, tests should be made.

Silt or Suspended Particles. Mixing water may contain up to 2,000 ppm of suspended clay or fine rock particles without adverse effect.

Oils. Mineral oil, not mixed with animal or vegetable oils, probably has less effect on strength development than other oils. However, mineral oil in concentrations greater than 2 percent by weight of concrete may reduce the concrete strength by more than 20 percent.

Algae. Mixing water containing algae may result in excessive reduction in concrete strength. Algae may also be present on aggregates, in which case the bond between the aggregate and the cement paste is weakened.

CONCRETE AGGREGATE

Even though aggregates are considered inert materials acting as filler, they make up from 60 to 80 percent of the volume of concrete. The characteristics of the aggregates have a considerable influence on the mix proportions and on the economy of the concrete. For example, rough-textured or flat and elongated particles require more water to produce workable concrete than do rounded or cubical particles. Hence, aggregate particles that are angular require more cement to maintain the same water-cement ratio and thus the concrete is more expensive. For most purposes, aggregates should consist of clean, hard, strong, durable particles free of chemicals or coatings of clay or other fine materials that affect the bond of the cement paste. Contaminating materials most often encountered are dirt, silt, clay, mica, salts, and humus or other organic matter that may appear as a coating or as loose, fine material. Many of them can be removed by washing but weak, friable or laminated aggregate particles are undesirable. Sand containing organic material cannot be washed clear. Shale or stones with shale and most cherts are especially undesirable. Visual inspection often discloses weaknesses in course aggregate. It should be tested in doubtful cases. The most commonly used aggregates are sand, gravel, crushed stone, and blast furnace slag. Cinders, burnt clay, expanded blast furnace slag, and other materials are also used. These aggregates produce normal-weight concrete, that is, concrete weighing from about 135 to 160 pounds per cubic foot. Normal-weight aggregates should meet the requirements of Specifications for Concrete Aggregates (ASTM C33). These specifications limit the permissible amounts of deleterious substances and cover requirements for gradation, abrasion resistance, and soundness. Aggregate characteristics, their significance, and standard tests for evaluation of these characteristics are given in table 2-1.

Abrasion Resistance. A general index of aggregate quality is the abrasion resistance of the aggregate. This characteristic is essential when the aggregate is used in concrete subject to abrasion as in heavy-duty floors.

Resistance to Freezing and Thawing. The freeze-thaw resistance of an aggregate is related to its porosity, absorption, and pore structure. This is an important characteristic in exposed concrete. If an aggregate particle absorbs so much

Characteristic	Significance or importance	Test or practice ASTM designation	Specification requirement
RESISTANCE TO ABRASION	Index of aggregate quality. Warehouse floors, loading platforms, pavements.	C131	Max. percent loss*
RESISTANCE TO FREEZING AND THAWING	Structures subjected to weathering.	C290, C291	Max. number of cycles
CHEMICAL STABILITY	Strength and durability of all types of structures.	C227 (mortar bar) C289 (chemical) C586 (aggregate prism) C295 (petrographic)	Max. expansion of mortar bar* Aggregates must not be re-active with cement alkalies*
PARTICLE SHAPE AND SURFACE TEXTURE	Workability of fresh concrete.		Max. percent flat and elongated pieces
GRADING	Workability of fresh concrete. Economy	C136	Max. and min. percent passing standard sieves
BULK UNIT WEIGHT	Mix design calculations. Classification	C29	Max. or min. unit weight (special concrete)
SPECIFIC GRAVITY	Mix design calculations.	C127 (coarse aggregate) C128 (fine aggregate)	
ABSORPTION AND SUR-FACE MOISTURE	Control of concrete quality.	C70, C127, C128	

*Aggregates not conforming to specification requirements may be used if service records or performance tests indicate they produce concrete having the desired properties.

Characteristics of Aggregates
Table 2-1

water that insufficient pore space is available, it will not accommodate water expansion that occurs during freezing. The performance of aggregates under exposure to freezing and thawing can be predicted in two ways: past performance and freezing-thawing tests of concrete specimens. If aggregates from the same source have previously given satisfactory service when used in concrete, the aggregate may be considered suitable. Aggregates not having a service record may be considered acceptable if they perform satisfactorily in concrete specimens subjected to freezing thawing tests (ASTM C290 and C291) and strength tests.

Chemical Stability. Aggregates which have chemical stability will neither react chemically with cement in a harmful manner nor be affected chemically by other external influences. Field service records generally provide the best information for the selection of nonreactive aggregates. If an aggregate has no service record and is suspected of being chemically unsound, laboratory tests are useful for determining its suitability. There are three ASTM tests for identifying alkali-reactive aggregates (ASTM C227, C289, and C586). In addition there is an ASTM recommended practice (C295) for testing of aggregates based on classification of the rock samples.

Particle Shape and Surface Texture. The particle shape and surface texture of an aggregate influence the properties of fresh concrete more than they affect the properties of hardened concrete. Very sharp and rough aggregate particles or flat, enlongated particles require more fine material to produce workable concrete than do aggregate particles that are more rounded or cubical. Stones which break up into long slivery pieces should be avoided or limited to about 15 percent in either fine or coarse aggregate.

Grading. The grading and maximum size of aggregate are important because of their relative effect on the workability, economy, porosity, and shrinkage of the concrete. Experience shows that either very fine or very coarse sands are objectionable; the first is uneconomical, the last gives harsh, unworkable mixes. The gradation or particle-size distribution of aggregate is determined by a sieve analysis. The standard sieves used for this purpose are numbers 4, 8, 16, 30, 50, and 100 for fine aggregate. The standard sieves used for the determination of the fineness modulus for coarse aggregate are 6-, 3-, 1½-, ¾-, and ⅜-inch and number 4. These sieves, for both fine and coarse aggregate grading, are based on square openings, the size of the openings in consecutive sieves being related by a constant ratio. Other sieves may be used for coarse aggregate grading, however. Grading charts, convenient for showing size distribution, generally have lines representing successive standard sieves placed at equal intervals as shown in figure 2–1. This figure also shows the grading limits for fine aggregates and for one designated size of coarse aggregate (ASTM C33). Fine aggregate is material which will pass a number 4 sieve and be retained on a number 100 sieve. Coarse aggregate is material retained by a number 4 sieve.

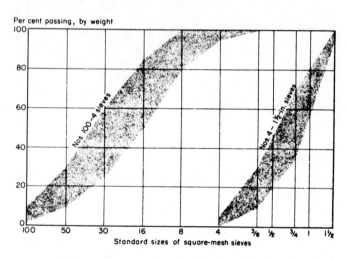

Per cent passing, by weight

Standard sizes of square-mesh sieves

Limits specified in ASTM C33 for fine aggregates and for one size of course aggregate
Figure 2-1

Fineness modulus. This is a term used as an index to the fineness or coarseness of aggregate. It is the summation of the cumulative percentages of the material retained on the standard sieves divided by 100. It is not an indication of grading, for an infinite number of gradings will give the same value for fineness modulus. To obtain the fineness modulus, quarter a sample of at least 500 grams of sand, and sieve through the No. 4, 8, 16, 30, 50, and 100 sieves. Weigh the material retained on each sieve and then calculate the cumulative weights retained. Determine the fineness modulus by adding up the cumulative percents and dividing by 100 (fig. 2–2). In general, fine aggregate with a very high or a very low value for fineness modulus will not be as good for concrete aggregate as medium sand. Coarse sand may not be workable, and fine sands are uneconomical. The aggregate fineness modulus from one source should not vary more than 0.20 from test samples taken at the source.

SCREEN SIZE	WEIGHT RETAINED (GRAMS)		CUMULATIVE % RETAINED
	INDIVIDUAL	CUMULATIVE	
NO. 4	40	40	40
NO. 8	130	170	17.0
NO. 16	130	300	30.0
NO. 30	250	550	55.0
NO. 50	270	820	82.0
NO. 100	100	920	92.0
PAN	80	--	--
TOTAL WEIGHT	1000	--	280.0
FINENESS MODULUS (FM)= $\frac{280}{100}$ = 2.80			

Typical calculation of fineness modules
Figure 2-2

Fine-aggregate grading. The most desirable fine aggregate grading depends on the type of work, richness of mix, and maximum size of coarse aggregate. In leaner mixes, or when small-size coarse aggregates are used, a grading that approaches the maximum recommended percentage passing each sieve is desirable for workability. In richer mixes, coarser gradings are desirable for economy. In general, if the water-cement ratio is kept constant and the ratio of fine to coarse aggregate is chosen correctly, a wide range in grading can be used without measurable effect on strength. The amount of fine aggregate passing the No. 50 and 100 sieves affects workability, finish and surface texture and water gain. For thin walls, hard-finished concrete floors, and smooth surfaces where concrete is cast against forms, the fine aggregate should contain not less than 15 percent passing the No. 50 sieve and at least 3 or 4 percent but not more than 10 percent passing the No. 100 sieve. With these minimum amounts of fines the concrete has better workability and is more cohesive so there is less water gain or bleeding than when lower percentages of fines are present. Aggregate gradings within the limits of ASTM C33 are generally satisfactory for most concretes.

Coarse-aggregate grading. The grading of a coarse aggregate of a given maximum size may be varied over a relatively wide range without appreciable effect on the cement and water requirements if the proportion of fine aggregate produces concrete of good workability. Table 2–2 indicates the gradation requirements for coarse

	Amount finer than each laboratory sieve (square openings), percent by weight							
Size Number	Nominal size (sieves) with square openings	4 in.	3½ in.	3 in.	2½ in.	2 in.	1½ in.	1 in.
1	3½ to 1½ in.	100	90 to 100		25 to 60		0 to 15	
2	2½ to 1½ in.			100	90 to 100	35 to 70	0 to 15	
357	2 in. to No. 4				100	95 to 100		35 to 70
467	1½ in. to No. 4					100	95 to 100	
57	1 in. to No. 4						100	95 to 100
67	¾ in. to No. 4							100
7	½ in. to No. 4							
8	⅜ in. to No. 8							
3	2 to 1 in.				100	90 to 100	35 to 70	0 to 15
4	1½ to ¾ in.					100	90 to 100	20 to 55

	Amount finer than each laboratory sieve (square openings), percent by weight						
Size Number	Nominal size (sieves) with square openings	¾ in.	½ in.	⅜ in.	No. 4 (4760-micron)	No. 8 (2380-micron)	No. 16 (1190-micron)
1	3½ to 1½ in.	0 to 5					
2	2½ to 1½ in.	0 to 5					
357	2 in. to No. 4		10 to 30		0 to 5		
467	1½ in. to No. 4	35 to 70		10 to 30	0 to 5		
57	1 in. to No. 4		25 to 100		0 to 10	0 to 5	
67	¾ in. to No. 4	90 to 100		20 to 55	0 to 10	0 to 5	
7	½ in. to No. 4	100	90 to 100	40 to 70	0 to 15	0 to 5	
8	⅜ in. to No. 8		100	85 to 100	10 to 30	0 to 10	0 to 5
3	2 to 1 in.		0 to 5				
4	1½ to ¾ in.	0 to 15		0 to 5			

*From specifications for concrete aggregate (ASTM–C33).

Gradation Requirements for Coarse Aggregate*
Table 2-2

aggregate. If wide variations occur in coarse aggregate grading, it may be necessary to vary the mix proportions in order to produce workable concrete. In such cases, it is often more economical to maintain grading uniformity in handling and manufacturing coarse aggregate than to adjust proportions for variations in gradation. Coarse aggregate should be graded up to the largest size that is practicable to use for the conditions of the job. The maximum size should not exceed ⅕ the dimension of nonreinforced members, ¾ of the clear spacing between reinforcing bars or between reinforcing bars and forms, and ⅓ the depth of nonreinforced slabs on ground. The larger the maximum size of the coarse aggregate, the less will be the mortar and paste necessary and the less water and cement required to produce a given quality. Field experience indicates that the amount of water required per unit volume of concrete for a given consistency and given aggregates is substantially constant regardless of the cement content or relative proportions of water to cement. Further, the water required decreases with increases in the maximum size of the aggregate. The water required per cubic yard of concrete with a slump of 3 to 4 inches is shown in figure 2–3 for a wide range in coarse aggregate sizes. It is apparent that, for a given water-cement ratio, the amount of cement required decreases (consequently, economy increases) as the maximum size of coarse aggregate increases. However, in some instances, especially in higher strength ranges, concrete with smaller maximum-size aggregate has a higher compressive strength than concrete with larger maximum-size aggregate at the same water-cement ratio.

Gap-graded aggregates. Certain particle sizes are lacking in gap-graded aggregates. Lack of two or more successive sizes may result in segregation problems, especially for non-air-entrained concretes with slumps greater than about 3 inches. If a stiff mix is required, gap-graded aggregates may produce higher strengths than normal aggregates used with comparable cement contents.

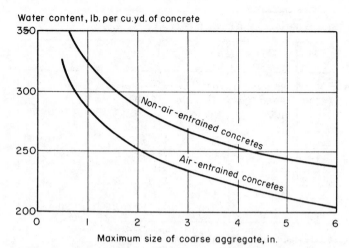

Water content, lb. per cu.yd. of concrete

Non-air-entrained concretes

Air-entrained concretes

Maximum size of coarse aggregate, in.

Water requirement for concrete of a given consistency
as a function of coarse aggregate size
Figure 2-3

Bulk Unit Weight. The bulk unit weight of an aggregate is the weight of the material used to fill a one-cubic-foot container. The term "bulk unit weight" is used since the volume contains both aggregate and voids. Methods of determining bulk unit weights of aggregates are given in ASTM C29.

Specific Gravity. The specific gravity of an aggregate is the ratio of its weight to the weight of an equal volume of water. Most normal-weight aggregates have specific gravities of 2.4 to 2.9. Methods of determining specific gravity for coarse and fine aggregates are given in ASTM C127. In concrete calculations, the specific gravities used are generally given for saturated, surface-dry aggregates; that is, all pores are filled with water, but no excess moisture is present on the surface. The internal structure of an aggregate particle is made up of solid matter and voids that may or may not contain water.

Absorption and Surface Moisture. Is is necessary to determine the absorption and surface moisture of aggregates so that the net water content of the concrete can be controlled and correct batch weights determined. The moisture conditions of aggregates are depicted in figure 2-4. The four states are—

(1) *Oven-dry*—pores bone-dry, fully absorbent.

(2) *Air-dry*—Dry at the surface but containing some interior moisture, thus somewhat absorbent.

(3) *Saturated surface-dry*—neither absorb-

ing water from nor contributing water to the concrete mix.

(4) *Damp or wet*—containing an excess of moisture on the surface. When fine aggregate is damp and is handled, bulking generally occurs. Bulking is the increase in volume caused by surface moisture holding the particles apart. The variation in the amount of bulking with the moisture content and grading is shown in figure 2–5. Since most sands are delivered in a damp condition, wide variations can occur in batch quantities if the batching is done by volume. Such variations are likely to be out of proportion to the moisture content of the sand. For this reason, proportioning by volume, if undertaken, should be done with considerable care.

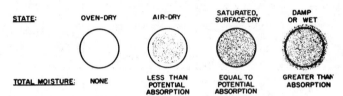

Moisture conditions of aggregates
Figure 2-4

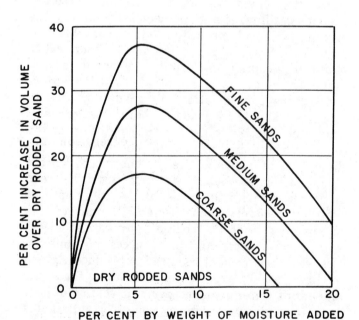

Variation in bulking with moisture and aggregate
grading
Figure 2-5

Harmful substances that may be present in aggregates include organic impurities, silt, clay, coal, lignite, and certain lightweight and soft particles. The effects of these substances on concrete and the ASTM test method designations are summarized in table 2–3.

Deleterious substances	Effect on concrete	Test ASTM designation
ORGANIC IMPURITIES	Affect setting and hardening, and may cause deterioration	C40 C87
MATERIALS FINER THAN NO. 200 SIEVE	Affect bond, and increase water requirement	C–117
COAL, LIGNITE, OR OTHER LIGHTWEIGHT MATERIALS	Affect durability, and may cause stains and popouts	C123
SOFT PARTICLES	Affect durability	C235
FRIABLE PARTICLES	Affect workability and durability, and may cause popouts	C142

Deleterious Substances in Aggregates
Table 2-3

HANDLING AGGREGATE

Aggregates should be handled and stored in such a manner as to minimize segregation and prevent contamination with deleterious substances. Aggregate is normally stored in stockpiles built up in layers of uniform thickness. Stockpiles should not be built up in high cone shapes nor allowed to run down slopes because this causes segregation. Aggregate should not be allowed to fall freely from the end of a conveyor belt. To minimize segregation, materials should be removed from stockpiles in approximately horizontal layers. If batching equipment is used, some of the aggregate will be stored in bins. Bins should be loaded by allowing the material to fall vertically over the outlet.

Chuting material at an angle against the side of the bin causes segregation of particles. Correct and incorrect methods of handling and storing aggregate are shown in figure 2–6.

FINE AGGREGATE STORAGE

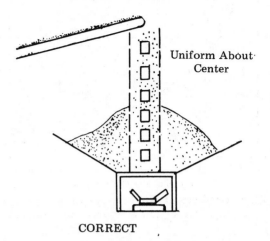

CORRECT

Chimney should surround material falling from and on conveyor, to prevent wind from separating fine and coarse materials. Openings should be provided as required to discharge materials at various elevations on the pile.

FINE AGGREGATE STORAGE

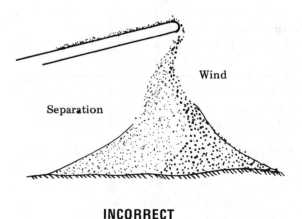

INCORRECT

Do not allow free fall of material from high end of conveyor which would permit wind to separate fine from coarse material.

Correct and incorrect handling and storing of aggregates
Figure 2-6

STORAGE IN BINS

CORRECT

Material is allowed to fall vertically over the outlet.

INCORRECT

Material is allowed to strike the side of the bin, causing segregation.

STOCKPILING OF COARSE AGGREGATE

PREFERABLE

A crane or other equipment should stockpile material in separate batches, each no larger than a truckload, so that it remains where placed and does not run down slopes.

FINISHED COARSE AGGREGATE STORAGE

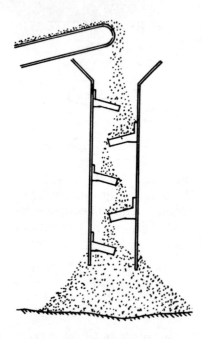

When stockpiling large sized aggregate from elevated conveyors, minimize breakage by use of a rock ladder.

Figure 2-6 - continued

LIMITED ACCEPTABILITY

Generally, a pile should not be built radially in horizontal layers by a bulldozer working with materials as dropped from a conveyor belt. A rock ladder may be needed in this ste-up.

Figure 2-6 - continued

12

CONCRETE ADMIXTURES

Admixtures include materials other than portland cement, water, and aggregates that are added to concrete, mortar, or grout immediately before or during mixing. Admixtures are sometimes used in concrete mixtures to improve certain quantities such as workability, strength, durability, watertightness, and wear resistance. They may also be added to reduce segregation, reduce heat of hydration, entrain air, and accelerate or retard setting and hardening. The same results can often be obtained by changing the mix proportions or by selecting other suitable materials without resort to admixtures (except air-entraining admixtures when necessary). Whenever possible, comparison should be made between these alternatives to determine which is more economical and/or convenient.

AIR-ENTRAINED CONCRETE

One of the advances in concrete technology in recent years has been the advent of air entrainment. The use of entrained air is recommended in concrete for most purposes. The principal reason for using intentionally entrained air is to improve concrete's resistance to freezing and thawing exposure. However, there are other important beneficial effects in both freshly mixed and hardened concrete. Air-entrained concrete is produced by using either an air-entraining cement or an air-entraining admixture during the mixing of the concrete. Unlike air entrapped in non-air-entrained concrete which exists in the form of relatively large air voids, which are not dispersed uniformly throughout the mix, entrained air exists in the form of minute disconnected bubbles well dispersed throughout the mass. These bubbles have diameters ranging from about one to three thousandths of an inch. As shown in figure 2–7, the bubbles are not interconnected and are well distributed throughout the paste. In general, air-entraining agents are derivatives of natural wood resins, animal or vegetable fats or oils, alkali salts of sulfated or sulfonated organic compounds and water-soluble soaps. Most air-entraining agents are in liquid form for use in the mix water. Instructions for the use of the various agents to produce a specified air content are provided by the manufacturer. Automatic dispensers made available by some manufacturers permit more accurate control of the quantities of air-entraining agents used in the mix.

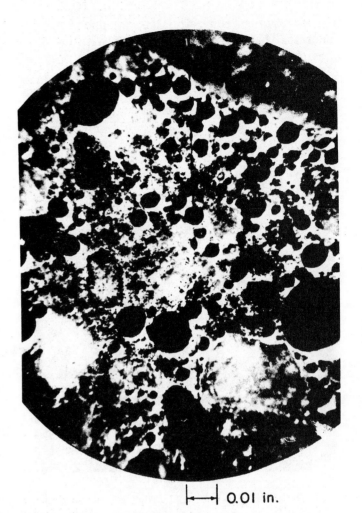

├──┤ 0.01 in.

Polished section of air-entrained concrete
magnified many times
Figure 2-7

Properties of Air-Entrained Concrete.

(1) *Workability.* Entrained air improves the workability of concrete. It is particularly effective in lean mixes and in mixes with angular and poorly graded aggregates. This improved workability allows a significant reduction in water and sand content. The disconnected air voids also reduce segregation and bleeding of plastic concrete.

(2) *Freeze-thaw resistance.* The freeze-thaw resistance of hardened concrete is significantly improved by the use of intentionally entrained air. As the water in concrete freezes, it expands, causing pressure that can rupture concrete. The entrained air voids act as reservoirs for excess water forced into them, thus relieving pressure and preventing damage to the concrete.

(3) *Resistance to de-icers.* Entrained air is effective for preventing scaling caused by de-icing chemicals used for snow and ice removal. The use

of air-entrained concrete is recommended for all concretes that come in any contact with de-icing chemicals.

(4) *Sulfate resistance.* The use of entrained air improves the sulfate resistance of concrete as shown in figure 2-8. Concrete made with a low water-cement ratio, entrained air, and cement having a low tricalcium aluminate content will be most resistant to attack from sulfate soil waters or seawater.

of non-air-entrained concrete of the same compressive strength. Abrasion resistance increases as the compressive strength increases.

(7) *Watertightness.* Air-entrained concrete is more watertight than non-air-entrained concrete, since entrained air inhibits the formation of inter-connected capillary channels. Air-entrained concrete should be used where watertightness is desired.

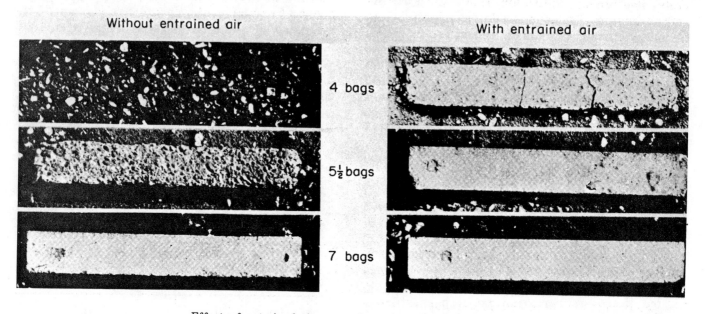

Effect of entrained air on performance of concrete specimens after 5 years of exposure to a sulfate soil

Figure 2-8

(5) *Strength.* Strength of air-entrained concrete depends principally upon the voids-cement ratio. For this ratio, "voids" is defined as the total volume of water plus air (entrained and entrapped). For a constant air content, strength varies inversely with the water-cement ratio. As air content is increased, a given strength generally may be maintained by holding to a constant voids-cement ratio by reducing amount of mixing water, increasing amount of cement, or both. Some reduction in strength may accompany air entrainment, but this is often minimized since air-entrained concretes have lower water-cement ratios than non-air-entrained concretes having the same slump. In some cases, however, it may be difficult to attain high strength with air-entrained concrete. Examples are when slumps are maintained constant as concrete temperatures rise and when certain aggregates are used.

(6) *Abrasion resistance.* Abrasion resistance of air-entrained concrete is about the same as that

Air-Entraining Materials. The entrainment of air in concrete may be accomplished by using air-entraining cement, by adding an air-entraining agent at the mixer or by a combination of both methods. Air-entraining cements should meet the specifications in ASTM C175. Commercial air-entraining admixtures, which are manufactured from a variety of materials, may be added at the mixer and should comply with the specifications in ASTM C260. Adequate control is required to ensure the proper air content at all times.

Factors Affecting Air Content.

(1) *Aggregate gradation and cement content.* Aggregate gradation and the cement content of a mix have a significant effect on the air content of both air-entrained and non-air-entrained concrete. For aggregate sizes smaller than 1½ inches, the air content increases sharply as the aggregate size decreases due to the increase in mortar volume. As the cement content increases, the air content de-

14

creases with the normal range of cement content.

(2) *Fine aggregate content.* The percentage of entrained air in concrete is affected by the fine aggregate content of the mix. Increasing the amount of fine aggregate causes more air to be entrained for a given amount of air-entraining cement or admixture.

(3) *Slump and vibration.* Slump and vibration affect the air content of air-entrained concrete. The greater the slump, the larger the percent reduction in air content during vibration. At all slumps, however, even 15 seconds of vibration causes a considerable reduction in air content. Even so, if vibration is properly applied, the air lost will consist mostly of large bubbles and little of the intentionally entrained air will be lost.

(4) *Concrete temperature.* Less air is entrained as the temperature of the concrete increases. The effect of temperature becomes more pronounced as the slump is increased.

(5) *Mixing action.* Mixing action is the most important factor in the production of entrained air in concrete. The amount of entrained air varies with the type and condition of the mixer, the amount of concrete being mixed, and the rate of mixing. Stationary mixers and transit mixers may produce concretes with significant differences in the amounts of air entrained. Mixers loaded to less than capacity may produce increases in air content, and decreases may result from overloading. Generally, more air is entrained as the speed of mixing is increased.

(6) *Admixtures and coloring agents.* Certain admixtures and coloring agents may reduce the amount of entrained air. This is particularly true of fly ash with high percentages of carbon. If calcium chloride is used, it should be added separately to the mix in solution form to prevent a chemical reaction with some air-entraining admixtures.

(7) *Premature finishing.* Premature finishing operations may cause excess water from bleeding to be worked into the top surface of concrete. The surface zones may become low in entrained air and thus susceptible to scaling.

Recommended Air Contents. The amount of air to be used in air-entrained concrete should be flexible and suited to the particular need. It should depend on the type of structure, climatic conditions, number of freeze-thaw-cycles, extent of exposure to de-icers and aggressive soils or waters,

and to some extent the strength of concrete. Field practice has shown that the amount of air indicated in table 2–4 should be specified for air-entrained concrete to provide adequate resistance to deterioration from freeze-thaw cycles and de-icing chemicals. Air contents are expressed in terms of percent by volume of the concrete, although air is entrained only in the mortar. In the field, a relatively constant amount of about 9 percent of air in the mortar fraction of the concrete (which corresponds to roughly 4½ percent by volume of the concrete) should provide the recommended air content for durability regardless of changes in cement content, maximum size of aggregate, consistency, and type of coarse aggregate. For exposure to extremely severe conditions, it is desirable to design air-entrained concrete for the highest air contents recommended in table 2–4. For certain concretes, such as those having high cement contents, low water contents, and consistencies below about one-inch slump, the level of air content recommended in table 2–4 is high. When entrained air is not required for protection against freeze-thaw or de-icers, the air content given in table 2–4 may be reduced by about one-third.

Maximum-size coarse aggregate, in.	Air content, percent by volume**
1½, 2, or 2½	5 ±1
¾ or 1	6 ±1
⅜ or ½ -----------	7½ ±1

*For structural lightweight concrete, add 2 percent to the values to allow for entrapped air in the aggregate particles, a range of ±1½ percent is permissible.
**The air content of the mortar fraction of the concrete should be about 9 percent.

Recommended Air Contents for Concretes Subject to Severe Exposure Conditions*
Table 2-4

Tests for Air Content. Available methods for determining air entrainment in freshly mixed concrete measure only air volume and not the air void characteristics. This has been shown to be generally indicative of the adequacy of the air void system when using air-entraining materials meeting ASTM specifications. Tests should be made regularly during construction with samples taken immediately after discharge from the mixer and also from the concrete after it has been placed and consolidated. The following methods for determining the air content of freshly mixed concrete have been standardized:

(1) *Pressure method.* This test (ASTM C231) is practical for field testing of all concretes

except those made with highly porous and light-weight aggregates.

(2) *Volumetric method.* This test (ASTM C173) is practical for field testing of all concretes, but is particularly useful for concretes made with lightweight and porous aggregates.

(3) *Gravimetric method.* This test (ASTM C138) is impractical as a field test method since it requires accurate knowledge of specific gravities and absolute volumes of concrete ingredients but can be satisfactorily used in the laboratory.

OTHER ADMIXTURES

Water-Reducing Admixtures. A water-reducing admixture is a material used for the purpose of reducing the quantity of mixing water required to produce concrete of a given consistency. These admixtures increase the slump for a given water content.

Retarding Admixtures. Retarders are sometimes used in concrete to reduce the rate of hydration to permit the placement and consolidation of concrete before the initial set. These admixtures are also used to offset the accelerating effect of hot weather on the setting of concrete. These admixtures generally fall in the categories of fatty acids, sugars, and starches.

Accelerating Admixtures. Accelerating admixtures accelerate the setting and the strength development of concrete. Calcium chloride is the most commonly used accelerator. It should be added in solution form as part of the mixing water and should not exceed 2 percent by weight of cement. Calcium chloride or other admixtures containing soluble chlorides should not be used in prestressed concrete, concrete containing embed-

ded aluminum, concrete in permanent contact with galvanized steel, or concrete subjected to alkali-aggregate reaction or exposed to soils or water containing sulfates.

Pozzolans. Pozzolans are siliceous or siliceous and aluminous materials which combine with calcium hydroxide to form compounds possessing cementitious properties. The properties of pozzolans and their effects on concrete vary considerably. Before one is used it should be tested in order to determine its suitability.

Workability Agents. It is often necessary to improve the workability of fresh concrete. Workability agents frequently used include entrained air, certain organic materials, and finely divided materials. Fly ash and natural pozzolans used should conform to ASTM C618.

Dampproofing and Permeability-Reducing Agents. Dampproofing admixtures, usually water-repellent materials, are sometimes used to reduce the capillary flow of moisture through concrete that is in contact with water or damp earth. Permeability-reducing agents are usually either water-repellents or pozzolans.

Grouting Agents. The properties of portland cement grouts are altered by the use of various air-entraining admixtures, accelerators, retarders, workability agents, and so on, in order to meet the needs of a specific application.

Gas-Forming Agents. Gas-forming materials may be added to concrete or grout in very small quantities to cause a slight expansion prior to hardening in certain applications. However, while hardening, the concrete or grout made with gas-forming material has a decrease in volume equal to or greater than that for normal concrete or grout.

CHAPTER 3

PROPORTIONING CONCRETE MIX

In arriving at proportional quantities of cement, water and aggregate for a concrete mix, one of three methods (book, trial batch, or absolute volume) is commonly used. The book method is a theoretical procedure in which established data is used to determine mix proportions. Due to the variation of the materials (aggregates) used, mixes arrived at by the book method require adjustment in the field following the mixing of trial batches and testing. Concrete mixtures should be designed to give the most economical and practical combination of the materials that will produce the necessary workability in the fresh concrete and the required qualities in the hardened concrete.

THE BOOK METHOD

Certain information must be known before a concrete mixture can be proportioned. The size and shape of structural members, the concrete strength required, and the exposure conditions must be determined. The water-cement ratio, aggregate characteristics, amount of entrained air, and slump are significant factors in the selection of the appropriate concrete mixture.

Water-Cement Ratio. In arriving at the water-cement ratio, the requirements of strength, durability, and watertightness of the hardened concrete must be considered. These factors are usually specified by the engineer in the design of the structure or assumed for purposes of arriving at tentative mix proportions. It is important to remember that a change in the water-cement ratio changes the characteristics of the hardened concrete. Selection of a suitable water-cement ratio is made from table 3-1 for various exposure conditions. Note that the quantities are the recommended maximum permissible water-cement ratios. As indicated in table 3-1, under certain conditions the water-cement ratio should be selected on the basis of concrete strength. In such cases, if possible, tests should be made with job materials to determine the relationship between water-cement ratio and strength. If laboratory test data or experience records for this relationship cannot be obtained, the necessary water-cement ratio may be estimated from figures 3-1 and 3-2, the lower edge of the applicable strength band curve should be used, and the desired design strength of the concrete should be increased by 15 percent according to ACI requirements. If flexural strength rather than compressive strength is the basis for design as in pavements, tests should be made to determine the relationship between water-cement ratio and flexural strength. An approximate relationship between flexural and compressive strength is—

$$f'_c = \left(\frac{R}{K} \right)^2$$

where f'_c = compressive strength, in psi

R = flexural strength (modulus of rupture), in psi, third-point loading

K = a constant, usually between 8 and 10.

In cases where both exposure conditions and strength must be considered, the lower of the two indicated water-cement ratios should be used.

Aggregate.

(1) *Fine aggregate.* Fine aggregate fills the spaces in the coarse aggregate and increases the workability of the mix. In general, aggregates which do not have a large deficiency or an excess of any size and give a smooth grading curve produce the most satisfactory mix. Fine aggregate grading and fineness modulus are discussed on page 8.

(2) *Coarse aggregate.* The largest size aggregate which is practical should be used. The larger the maximum size of the coarse aggregate, the less mortar and paste will be necessary. It follows that the larger the coarse aggregate, the less water and cement will be required for a given quality of concrete. The maximum size aggregate should not exceed one-fifth the minimum dimension of the member, or three-fourths of the clear space between reinforcing bars. For pavement or floor slabs, the maximum size aggregate should not exceed one-third the slab thickness. The maximum size of coarse aggregate that produces concrete of maximum strength for a given cement content depends upon aggregate source as well as aggregate shape and grading. For many aggregates, this "optimum" maximum size is about ¾ inch.

Type of structures	Severe wide range in temperature or frequent alternations of freezing and thawing (air-entrained concrete only) (gallons/sack)			Mild temperature rarely below freezing, or rainy, or arid (gallons/sack)		
	In air	At water line or within range of fluctuating water level or spray		In air	At water line or within range of fluctuating water level or spray	
		In fresh water	In sea water or in contact with sulfates†		In fresh water	In sea water or in contact with sulfates†
A. Thin sections such as reinforced piles and pipe	5.5	5	4.5	6	5.5	4.5
B. Bridge decks	5	5	4.5	5.5	5.5	5
C. Thin sections such as railings, curbs, sills, ledges, ornamental or architectural concrete, and all sections with less than 1-in. concrete cover over reinforcement	5.5			6	5.5	
D. Moderate sections, such as retaining walls, abutments, piers, girders, beams	6	5.5	5	††	6	5
E. Exterior portions of heavy (mass) sections	6.5	5.5	5	††	6	5
F. Concrete deposited by tremie under water		5	5		5	5
G. Concrete slabs laid on the ground	6			††		
H. Pavements	5.5			6		
I. Concrete protected from the weather, interiors of buildings, concrete below ground	††			††		
J. Concrete which will later be protected by enclosure or backfill but which may be exposed to freezing and thawing for several years before such protection is offered	6			††		

*Adapted from Recommended Practice for Selecting Proportions for Concrete (ACI 613–54).
**Air-entrained concrete should be used under all conditions involving severe exposure and may be used under mild exposure conditions to improve workability of the mixture.
†Soil or groundwater containing sulfate concentrations of more than 0.2 per cent. For moderate sulfate resistance, the tricalcium aluminate content of the cement should be limited to 8 per cent, and for high-sulfate resistance to 5 per cent. At equal cement contents, air-entrained concrete is significantly more resistant to sulfate attack than non-air-entrained concrete.
††Water-cement ratio should be selected on basis of strength and workability requirements, but minimum cement content should not be less than 470 lb. per cubic yard.

ACI Recommended Maximum Permissible Water-Cement Ratios for Different
Types of Structures and Degrees of Exposure*
Table 3-1

WATER - U. S. GAL. PER SACK OF CEMENT

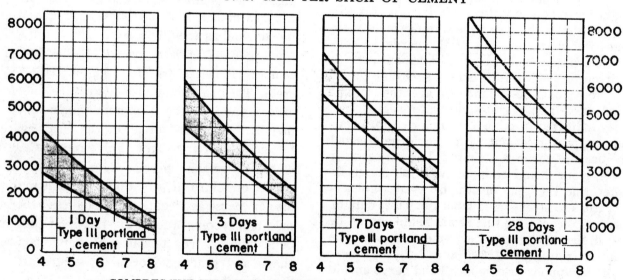

COMPRESSIVE STRENGTH - LBS PER SQ. IN. MOIST CURED AT 70° F
Age-compressive strength relationships for types I and III non-air-entrained portland cement
Figure 3-1

WATER - U.S. GAL. PER SACK OF CEMENT

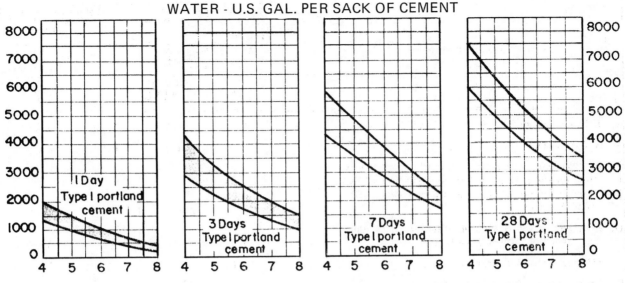

COMPRESSIVE STRENGTH - LBS PER SQ . IN. MOIST CURED AT 70° F
Age-compressive strength relationships for types I and III non-air-
entrained portland cement - continued
Figure 3-1

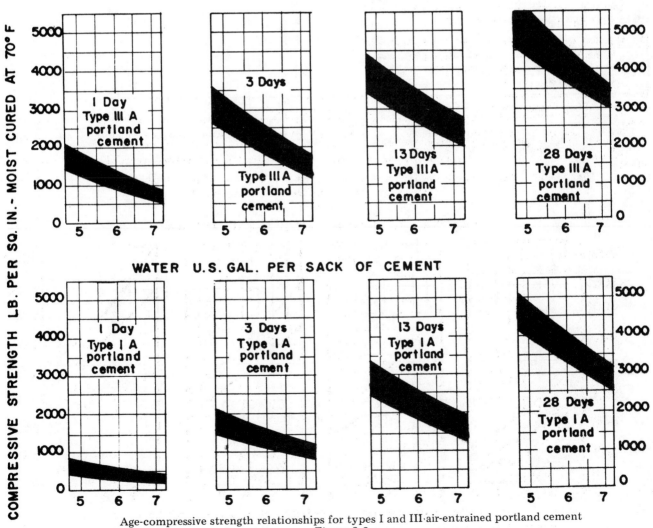

Age-compressive strength relationships for types I and III air-entrained portland cement
Figure 3-2

19

Water-cement ratio Gal per sack	Maximum size of aggregate inches	Air content (entrapped air) per cent	Water gal per cu yd of concrete	Cement sacks per cu yd of concrete	With fine sand—fineness modulus = 2.50		
					Fine aggregate per cent of total aggregate	Fine aggregate lb per cu yd of concrete	Coarse aggregate lb per cu yd of concrete
4.5	3/8	3	46	10.3	50	1240	1260
	1/2	2.5	44	9.8	42	1100	1520
	3/4	2	41	9.1	35	960	1800
	1	1.5	39	8.7	32	910	1940
	1 1/2	1	36	8.0	29	880	2110
5.0	3/8	3	46	9.2	51	1330	1260
	1/2	2.5	44	8.8	44	1180	1520
	3/4	2	41	8.2	37	1040	1800
	1	1.5	39	7.8	34	990	1940
	1 1/2	1	36	7.2	31	960	2110
5.5	3/8	3	46	8.4	52	1390	1260
	1/2	2.5	44	8.0	45	1240	1520
	3/4	2	41	7.5	38	1090	1800
	1	1.5	39	7.1	35	1040	1940
	1 1/2	1	36	6.5	32	1000	2110
6.0	3/8	3	46	7.7	53	1440	1260
	1/2	2.5	44	7.3	46	1290	1520
	3/4	2	41	6.8	39	1130	1800
	1	1.5	39	6.5	36	1080	1940
	1 1/2	1	36	6.0	33	1040	2110
6.5	3/8	3	46	7.1	54	1480	1260
	1/2	2.5	44	6.8	46	1320	1520
	3/4	2	41	6.3	39	1190	1800
	1	1.5	39	6.0	37	1120	1940
	1 1/2	1	36	5.5	34	1070	2110
7.0	3/8	3	46	6.6	55	1520	1260
	1/2	2.5	44	6.3	47	1360	1520
	3/4	2	41	5.9	40	1200	1800
	1	1.5	39	5.6	37	1150	1940
	1 1/2	1	36	5.1	34	1100	2110
7.5	3/8	3	46	6.1	55	1560	1260
	1/2	2.5	44	5.9	48	1400	1520
	3/4	2	41	5.5	41	1240	1800
	1	1.5	39	5.2	38	1190	1940
	1 1/2	1	36	4.8	35	1130	2110
8.0	3/8	3	46	5.7	56	1600	1260
	1/2	2.5	44	5.5	48	1440	1520
	3/4	2	41	5.1	42	1280	1800
	1	1.5	39	4.9	39	1220	1940
	1 1/2	1	36	4.5	35	1160	2110

*See footnote at end of table.

Suggested Trial Mixes for Non-Air-Entrained Concrete of Medium Consistency
with 3- to 4-Inch Slump*
Table 3-4

With average sand—fineness modulus = 2.75			With coarse sand—fineness modulus = 2.90		
Fine aggregate percent of total aggregate	Fine aggregate lb per cu yd of concrete	Coarse aggregate lb per cu yd of concrete	Fine aggregate percent of total aggregate	Fine aggregate lb per cu yd of concrete	Coarse aggregate lb per cu yd of concrete
52	1310	1190	54	1350	1150
45	1170	1450	47	1220	1400
37	1030	1730	39	1080	1680
34	980	1870	36	1020	1830
32	960	2030	33	1000	1990
54	1400	1190	56	1440	1150
46	1250	1450	48	1300	1400
39	1110	1730	41	1160	1680
36	1060	1870	38	1100	1830
34	1040	2030	35	1080	1990
55	1460	1190	57	1500	1150
47	1310	1450	49	1360	1400
40	1160	1730	42	1210	1680
37	1110	1870	39	1150	1830
35	1080	2030	36	1120	1990
56	1510	1190	57	1550	1150
48	1360	1450	50	1410	1400
41	1200	1730	43	1250	1600
38	1150	1870	39	1190	1830
36	1120	2030	37	1160	1990
57	1550	1190	58	1590	1150
49	1390	1450	51	1440	1400
42	1240	1730	43	1290	1680
39	1190	1870	40	1230	1830
36	1150	2030	37	1190	1990
57	1590	1190	59	1630	1150
50	1430	1450	51	1480	1400
42	1270	1730	44	1320	1680
39	1220	1870	41	1260	1830
37	1180	2030	38	1220	1990
58	1630	1190	59	1670	1150
50	1470	1450	52	1520	1400
43	1310	1730	45	1370	1600
40	1260	1870	42	1300	1830
37	1210	2030	39	1250	1990
58	1670	1190	60	1710	1150
51	1520	1450	53	1560	1400
44	1350	1730	45	1400	1680
41	1290	1870	42	1330	1830
38	1250	2030	39	1280	1990

*Increase or decrease water per cubic yard by 3 per cent for each increase or decrease of 1 in. in slump, then calculate quantities by absolute volume method. For manufactured fine aggregate, increase percentage of fine aggregate by 3 and water by 17 lb. per cubic yard of concrete. For less workable concrete, as in pavements, decrease percentage of fine aggregate by 3 and water by 8 lb. per cubic yard of concrete.

Suggested Trial Mixes for Non-Air-Entrained Concrete of Medium Consistency
with 3- to 4-Inch Slump* —Continued
Table 3-4

Water-cement ratio Gal per sack	Maximum size of aggregate inches	Air Content (entrapped air) per cent	Water gal per cu yd of concrete	Cement sacks per cu yd of concrete	With fine sand—fineness modulus = 2.50		
					Fine aggregate per cent of total aggregate	Fine aggregate lb per cu yd of concrete	Coarse aggregate lb per cu yd of concrete
4.5	⅜	7.5	41	9.1	50	1250	1260
	½	7.5	39	8.7	41	1060	1520
	¾	6	36	8.0	35	970	1800
	1	6	34	7.8	32	900	1940
	1½	5	32	7.1	29	870	2110
5.0	⅜	7.5	41	8.2	51	1330	1260
	½	7.5	39	7.8	43	1140	1520
	¾	6	36	7.2	37	1040	1800
	1	6	34	6.8	33	970	1940
	1½	5	32	6.4	31	930	2110
5.5	⅜	7.5	41	7.5	52	1390	1260
	½	7.5	39	7.1	44	1190	1520
	¾	6	36	6.5	38	1090	1800
	1	6	34	6.2	34	1010	1940
	1½	5	32	5.8	32	970	2110
6.0	⅜	7.5	41	6.8	53	1430	1260
	½	7.5	39	6.5	45	1230	1520
	¾	6	36	6.0	38	1120	1800
	1	6	34	5.7	35	1040	1940
	1½	5	32	5.3	32	1010	2110
6.5	⅜	7.5	41	6.3	54	1460	1260
	½	7.5	39	6.0	45	1260	1520
	¾	6	36	5.5	39	1150	1800
	1	6	34	5.2	36	1080	1940
	1½	5	32	4.9	33	1040	2110
7.0	⅜	7.5	41	5.9	54	1500	1260
	½	7.5	39	5 6	46	1300	1520
	¾	6	36	5.1	40	1180	1800
	1	6	34	4.9	36	1100	1940
	1½	5	32	4.6	33	1060	2110
7.5	⅜	7.5	41	5.5	55	1530	1260
	½	7.5	39	5.2	47	1330	1520
	¾	6	36	4.8	40	1210	1800
	1	6	34	4.5	37	1140	1940
	1½	5	32	4.3	34	1090	2110
8.0	⅜	7.5	41	5.1	55	1560	1260
	½	7.5	39	4.9	47	1360	1520
	¾	6	36	4.5	41	1240	1800
	1	6	34	4.3	37	1160	1940
	1½	5	32	4.0	34	1110	2110

*See footnote at end of table.

Suggested Trial Mixes for Air-Entrained Concrete of Medium Consistency with 3- to 4-Inch Slump*
Table 3-5

With average sand—fineness modulus = 2.75			With coarse sand—fineness modulus = 2.90		
Fine aggregate percent of total aggregate	Fine aggregate lb per cu yd of concrete	Coarse aggregate lb per cu yd of concrete	Fine aggregate percent of total aggregate	Fine aggregate lb per cu yd of concrete	Coarse aggregate lb per cu yd of concrete
53	1320	1190	54	1360	1150
44	1130	1450	46	1180	1400
38	1040	1730	39	1090	1680
34	970	1870	36	1010	1830
32	950	2030	33	990	1990
54	1400	1190	56	1440	1150
46	1210	1450	47	1260	14000
39	1110	1730	41	1160	1630
36	1040	1870	37	1080	1830
33	1010	2030	35	1050	1990
55	1460	1190	57	1500	1150
46	1260	1450	48	1310	1400
40	1160	1730	42	1210	1680
37	1080	1870	38	1120	1830
34	1050	2030	35	1090	1990
56	1500	1190	57	1540	1150
47	1300	1450	49	1350	1400
41	1190	1730	42	1240	1680
37	1110	1870	39	1150	1830
35	1090	2030	36	1130	1990
56	1530	1190	58	1570	1150
48	1330	1450	50	1380	1400
41	1220	1730	43	1270	1680
38	1150	1870	39	1190	1830
36	1120	2030	37	1160	1990
57	1570	1190	58	1610	1150
49	1370	1450	50	1420	1400
42	1250	1730	44	1300	1680
38	1170	1870	40	1210	1830
36	1140	2030	37	1180	1990
57	1600	1190	59	1640	1150
49	1400	1450	51	1450	1400
43	1280	1730	44	1330	1680
39	1210	1870	41	1250	1830
37	1170	2030	38	1210	1990
58	1630	1190	59	1670	1150
50	1430	1450	51	1480	1400
43	1310	1730	44	1360	1680
40	1230	1870	41	1270	1830
37	1190	2030	38	1230	1990

*Increase or decrease water per cubic yard by 3 per cent for each increase or decrease of 1 in. in slump, then calculate quantities by absolute volume method For manufactured fine aggregate, increase percentage of fine aggregate by 3 and water by 17 lb. per cubic yard of concrete. For less workable concrete, as in pavements, decrease percentage of fine aggregate by 3 and water by 8 lb. per cubic yard of concrete.

Suggested Trial Mixes for Air-Entrained Concrete of Medium Consistency with 3- to 4-Inch Slump* —Continued
Table 3-5

Entrained Air. Entrained air should be used in all concrete exposed to freezing and thawing and may be used for mild exposure conditions to improve workability. It is recommended for all paving concrete regardless of climatic conditions. The recommended total air contents for air-entrained concretes are shown in table 3–2. When mixing water is held constant, the entrainment of air will increase slump. When cement content and slump are held constant, less mixing water is required; the resulting decrease in the water-cement ratio helps to offset possible strength decreases and results in improvements in other paste properties such as permeability. Hence, the strength of air-entrained concrete may equal, or nearly equal, that of non-air-entrained concrete when their cement contents and slumps are the same.

Type of construction	Slump, inches	
	Maximum	Minimum
Reinforced foundation walls and footings__	6	3
Unreinforced footings, caissons, and substructure walls_____	4	1
Reinforced slabs, beams, and walls_____	6	3
Building columns_____	6	4
Pavements_____	3	1
Heavy mass construction_____	3	1
Bridge decks_____	4	3
Sidewalk, driveway, and slabs on ground__	6	3

*When high-frequency vibrators are used, the values may be decreased approximately one-third, but in no case should the slump exceed 6 inches.

Recommended Slumps for Various Types of Construction*
Table 3-3

Maximum size of aggregate, in.	Air-entrained concrete				Non-air-entrained concrete			
	Recommended average total air content, per cent†	Slump, in.			Approximate amount of entrapped air, per cent	Slump, in.		
		1 to 2	3 to 4	5 to 6		1 to 2	3 to 4	5 to 6
		Water, gal. per cu.yd. of concrete**				Water, gal. per cu.yd. of concrete**		
3/8	7.5	37	41	43	3.0	42	46	49
1/2	7.5	36	39	41	2.5	40	44	46
3/4	6.0	33	36	38	2.0	37	41	43
1	6.0	31	34	36	1.5	36	39	41
1 1/2	5.0	29	32	34	1.0	33	36	38
2	5.0	27	30	32	0.5	31	34	36
3	4.0	25	28	30	0.3	29	32	34
6	3.0	22	24	26	0.2	25	28	30

*Adapted from Recommended Practice for Selecting Proportions for Concrete (ACI 613–54).
**These quantities of mixing water are for use in computing cement factors for trial batches. They are maximums for reasonably well-shaped angular coarse aggregates graded within limits of accepted specifications.
†Plus or minus 1 per cent.

Approximate Mixing Water Requirements for Different Slumps and Maximum Sizes of Aggregates*
Table 3-2

Slump. The slump test is generally used as a measure of the consistency of concrete. It should not be used to compare mixes with wholly different proportions or mixes with different kinds of sizes of aggregates. When used to test different batches of the same mixture, changes in slump indicate changes in materials, mix proportions, or water content. Acceptable slump ranges are indicated in table 3–3.

Mix Proportions. Knowing the water-cement ratio, slump, maximum size of aggregate, and fineness modulus, tables of trial mixes such as tables 3–4 and 3–5 can be used to determine the proportions of trial mixes. The quantities in tables 3–4 and 3–5 are based on concrete having a slump of 3 to 4 inches, with well graded aggregates having a specific gravity of 2.65. For other conditions it is necessary to adjust the quantities in accordance with the footnotes.

Example. The mix proportions are to be determined for a water-cement ratio of 6 gallons per sack, maximum aggregate size of 1 inch, air content of 6 ± 1 percent, and a slump of 3 to 4 inches. The fine aggregate has a fineness modulus of 2.50 and a moisture content of 5 percent. The coarse aggregate has a moisture content of 1 percent.

Table 3–5 for air-entrained concrete will be used because of the air content requirement. The following quantities per cubic yard are taken from table 3–5:

Cement	5.7 sacks =	535 pounds
Water	34 gallons =	285 pounds
Fine aggregate		1,040 pounds
Coarse aggregate		1,940 pounds
Total		3,800 pounds

The moisture content of the aggregates must be considered since the tables are for the saturated, surface-dry condition. The free moisture in the fine aggregate is 5 percent of 1,040 pounds or 52 pounds (use 50 pounds). The free moisture in the coarse aggregate is 1 percent of 1,940 pounds or approximately 19 pounds (use 20 pounds). The corrected weights per cubic yard of concrete are—

Cement		535 pounds
Water	285 — 50 — 20 =	215 pounds
Fine aggregate	1,040 + 50 =	1,090 pounds
Coarse aggregate	1,940 + 20 =	1,960 pounds
Total		3,800 pounds

When these quantities were mixed the consistency was such that the slump was approximately 1 inch. Additional water (25 pounds) was added to bring the slump to slightly more than 3 inches. The unit weight was measured to be 145 pounds per cubic foot. Since the unit weight of the concrete was 145 pounds per cubic foot and the total weight of concrete was 3,800 + 25 = 3,825 pounds, the volume of concrete was 26.4 cubic feet. This is slightly less than 1 cubic yard (27 cubic feet). The principal reason for this discrepancy is that the specific gravity values of the aggregates were probably different from those assumed in tables 3–4 and 3–5. In small adjustments it may be assumed that the unit weight of the concrete remains essentially constant and that the amount of water required per cubic yard of concrete remains constant. The adjusted water requirement is—

$$\frac{27}{26.4} \times 310 \text{ pounds} = 315 \text{ pounds} = 38 \text{ gallons}$$

Note that the 310 pounds used is the total amount of water needed, 285 + 25 pounds. The adjusted cement requirement is

$$\frac{38 \text{ gallons}}{\frac{6 \text{ gallons}}{\text{sack}}} = 6.3 \text{ sacks} = 595 \text{ pounds}$$

The weight of materials per cubic yard of concrete must total 145 × 27 = 3,910 pounds, approximately. The total weight of aggregates must therefore be 3,910 — 315 — 595 = 3,000 pounds. Table 3–5 indicates that 35 percent of this should be fine aggregate.

The adjusted trial mix proportions (per cubic yard) are therefore:

Cement	6.3	sacks
Water	38	gallons
Fine aggregate	1,050	pounds
Coarse aggregate	1,950	pounds

TRIAL BATCH METHOD

The trial batch method of mix design utilizes actual materials in arriving at mix proportions instead of the tables of trial mixes (tables 3–4 and 3–5). When the quality of the concrete mixture is specified in terms of the water-cement ratio, the trial batch procedure consists essentially of combining a paste (water, cement, and, generally, entrained air) of the correct proportions with the necessary amounts of fine and coarse aggregates to produce the required slump and workability. Quantities per sack and/or per cubic yard are then calculated. It is important to use representative samples of the aggregates, cement, water, and air-entraining admixture, if used. The aggregates should be pre-wetted; allowed to dry to a saturated, surface-dry condition; and placed in covered containers to keep them in this condition until use. This procedure simplifies calculations and eliminates error caused by variations in aggregate moisture content. The size of the trial

CONCRETE

TRIAL MIX DATA

1. PROJECT NO: _____

2. STRUCTURE: *Retaining Wall*

3. EXPOSURE CONDITION:
   Severe or Moderate ✔ Mild ____

   In air ____
   In Fresh Water ✔
   In Sea Water ____

4. TYPE OF STRUCTURE (A-I): *D*

   MAX. W/C FOR EXPOSURE: *5.5* gal/sack

   MAX. W/C FOR WATERTIGHTNESS *55* gal/sack

5. TYPE OF CEMENT *1A*

6. FINENESS MODULUS OF SAND *2.75*

7. SPECIFIC GRAVITY:
   Sand *2.60*
   Gravel *2.65*

8. MAXIMUM SIZE AGGREGATE: *1½"*

9. AIR CONTENT *5* % ± 1%

10. DESIRED SLUMP RANGE
    MAX. *4* in.   MIN. *2* in.

11. STRENGTH REQUIREMENT: *3450* psi

    W/C FOR STRENGTH: *5.75* gal/bag

    USE W/C *5.5* gal/sack

    gal/sack

DATA FOR TRIAL BATCH
(Saturated, surface-dry aggregates)

(1) Material	(2) Initial Wt. (lb)	(3) Final Wt. (lb)	(4) Wt. Used (lb)	(5) Wt. for 1-bag Batch	(6) Wt. per cu.yd.	(7) REMARKS
Cement	20.0	0	20.0	94.0	540	Cement Factor = 5.74 bag/cu.yd.
Water	9.8	0	9.8	46.1	265	31.8 gal/cu.yd.
Fine Agg.	66.2	27.9	38.3	180.0	1,035	33.5 % of total aggregate
Coarse Agg.	89.8	13.8	76.0	357.0	2,050	
Air-Entraining Admixture	.30 cc		TOTAL (T) = 677.1			

Measured Slump: *3* in.   Air Content *5.4* %   Workability: *Good*

Wt. Container concrete (lb.)	42.6
Wt. Container (lb.)	6.6
Wt. Concrete = A (lb.)	36.0
Vol. Container = B (cu.ft.)	.25

Unit Wt. of concrete = $w = \dfrac{A}{B}$

$\dfrac{A}{B} = \dfrac{36.0}{.25} = 144$ lb/cu.ft.

Yield $= \dfrac{T}{w} = \dfrac{677.1}{144} = 4.7$ cu.ft./bag

Worksheet for concrete trial mix data
Figure 3-3

CALCULATIONS

Column (4) = Column (2) minus Column (3)

Column (5) = Column (4) times (94 ÷ Wt. of Cement Used)

$$\text{Yield} = \frac{\text{Total Wt. of material for 1-bag batch (T)}}{\text{Unit wt. of concrete (w)}}$$

Cement factor = 27 ÷ yield

Column (6) = Column (5) times cement factor

$$\text{Gal./cu.yd.} = \frac{(\text{Cement factor}) \times (\text{wt. of water for 1-bag batch})}{8.33 \text{ (lb. of water per gal.)}}$$

$$= (\text{Cement factor}) \times (W/C)$$

Figure 3–3—Continued

batch is dependent on the equipment and the test specimens to be made. Batches using 10 to 20 pounds of cement may be adequate, although larger batches will produce more accurate data. Machine mixing is recommended since it more nearly represents job conditions; it is mandatory if the concrete is to contain entrained air.

The mix proportions are to be determined for concrete which will be used in a retaining wall that will be exposed to fresh water in a severe climate. A compressive strength of 3,000 psi at 28 days is required. The minimum thickness of the wall is 8 inches and 2 inches of concrete must cover the reinforcement. All trial mix data will be entered in the appropriate blanks on the trial mix data worksheet, figure 3–3.

Line D of table 3-1 indicates a maximum water-cement ratio of 5.5 gallons per sack will satisfy the exposure requirements. Using type IA (air-entrained) portland cement and a compressive strength of 3,450 psi (3,000 psi +15 percent), figure 3–2 indicates that a maximum water-cement ratio of approximately 5.75 gallons per sack will satisfy the strength requirements. In order to meet both specifications a water-cement ratio of 5.5 gallons per sack is selected. Since the maximum size of coarse aggregate must not exceed one-fifth the minimum thickness of the wall, nor three-fourths of the clear space between reinforcement and the surfaces, the maximum size of coarse aggregate is chosen as 1½ inches. Because

of the severe exposure conditions, the concrete should contain entrained air. From table 3–2, the recommended air content is 5 ± 1 percent. If we assume that the concrete will be consolidated by vibration, table 3–3 indicates a recommended slump of from 2 to 4 inches. The trial batch proportions are now determined. A batch containing 20 pounds of cement is chosen for convenience. The mixing water required is therefore

$$\frac{20}{94} \times 5.5 \frac{\text{gal}}{\text{sack}} \times 8.33 \frac{\text{lb}}{\text{gal}} = 9.8 \text{ pounds}$$

Representative samples of fine and coarse aggregates are selected and weighed. This is recorded in column (2) of figure 3–3. All of the measured quantities of cement, water, and air-entraining admixture are used. Fine and coarse aggregates are then added until a workable mixture having the proper slump is produced. Figure 3–4 indicates the appearance of fresh concrete with correct and incorrect amounts of mortar. The weight of material actually used is recorded in column (4). The weights for a 1-bag batch and per cubic yard are calculated and recorded in columns (5) and (6), respectively. The cement factor in bags per cubic yard is calculated and recorded as indicated in figure 3–3. The percentage of fine aggregate by weight of total aggregate is also included as is the yield of concrete in cubic feet per bag. The slump, air content, workability, and unit weight of concrete are determined and noted as

27

(a) A concrete mixture in which there is not sufficient cement-sand mortar to fill all the spaces between coarse aggregate particles. Such a mixture will be difficult to handle and place and will result in rough honeycombed surfaces and porous concrete.

(b) A concrete mixture which contains correct amount of cement-sand mortar. With light troweling all spaces between coarse aggregate particles are filled with mortar. Note appearance on edges of pile. This is a good workable mixture and will give maximum yield of concrete with a given amount of cement.

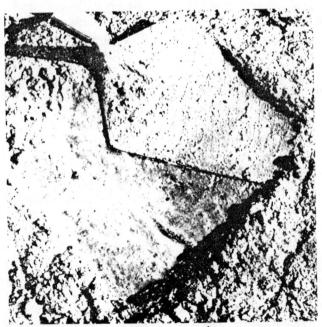

(c) A concrete mixture in which there is an excess of cement-sand mortar. While such a mixture is plastic and workable and will produce smooth surfaces, the yield of concrete will be low and consequently uneconomical. Such concrete is also likely to be porous.

Appearance of concrete mixes with correct and incorrect amounts of mortar
Figure 3-4

shown. To determine the most economical proportions, additional trial batches should be made varying the percentage of fine aggregate. In each batch the water-cement ratio, aggregate gradations, air content, and slump are maintained approximately the same. Results of four such trial batches are summarized in table 3–6. For these mixes, the percentage of fine aggregate is plotted against the cement factor in figure 3–5. The minimum cement factor (5.72 sacks per cubic yard, use 5.7) occurs at a fine aggregate content of about 32 percent of total aggregate. Since the water-cement ratio is 5.5 gallons per sack and the unit weight of the concrete for an air content of 5 percent is about 144 pounds per cubic foot, the final quantities for the mix proportion (per cubic yard) are—

Cement	5.7 sacks	= 535 pounds
Water 5.7 sacks × 5.5 gal/sack 31.5 gallons		= 260 pounds
	Total	795 pounds
Concrete per cubic yard 144 x 27	=	3,890 pounds
Aggregates 3,890 — 795	=	3,095 pounds
Fine aggregate .32 × 3,095	=	990 pounds
Coarse aggregate 3,095 — 990	=	2,105 pounds

Batch no.	Slump, in.	Air content, per cent	Unit wt., pcf	Cement factor, sacks per cu. yd.	Fine aggregate, per cent of total aggregate	Work-ability
1	3	5.4	144	5.74	33.5	Excellent
2	2¾	4.9	144	5.91	27.4	Harsh
3	2½	5.1	144	5.84	35.5	Excellent
4	3¼	4.7	145	5.74	30.5	Good

*The water-cement ratio selected was 5.5 gal/sack.

Examples of Results of Laboratory Trial Mixes*
Table 3-6

ABSOLUTE VOLUME METHOD

Concrete mixtures may be proportioned using absolute volumes. This method is detailed in the American Concrete Institute (ACI) report, Recommended Practice for Selecting Proportions for Concrete (ACI 613–54). In this method, the water-cement ratio, slump, air content, and maximum size of aggregate are selected as before. In addition, the water requirement is estimated from table 3–2. Certain additional items must be known before calculations can be made. These are: the specific gravities of fine and coarse aggregates, the dry-rodded unit weight of coarse aggregate, and the fineness modulus of the fine aggregate. If the

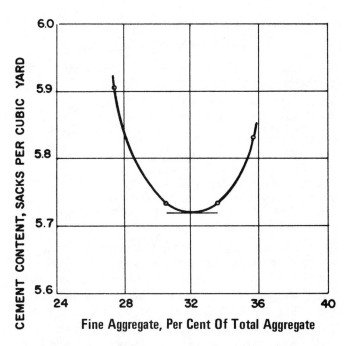

Relationship between percentage of fine aggregate and cement content for a given water-cement ratio and slump
Figure 3-5

maximum size of aggregate and the fineness modulus of the fine aggregate are known, the volume of dry-rodded coarse aggregate per cubic yard can be estimated from table 3–7. Then the quantities per cubic yard of water, cement, coarse aggregate, and air can be calculated. The sum of the absolute volumes of these materials in cubic feet is then subtracted from 27 to give the specific volume of fine aggregate.

Maximum size of aggregate, in.	Fineness modulus of fine aggregate			
	2.40	2.60	2.80	3.00
	Coarse aggregate, cu. ft. per cu. yd.*			
⅜	13.5	13.0	12.4	11.9
½	15.9	15.4	14.8	14.3
¾	17.8	17.3	16.7	16.2
1	19.2	18.6	18.1	17.6
1½	20.2	19.7	19.2	18.6
2	21.1	20.5	20.0	19.4
3	22.1	21.6	21.1	20.5

*Volumes are based on aggregates in dry-rodded condition as described in Method of Test for Unit Weight of Aggregate (ASTM C29). These volumes are selected from empirical relationships to produce concrete with a degree of workability suitable for usual reinforced construction. For less workable concrete such as required for concrete pavement construction, they may be increased about 10 percent. When placement is to be by pump, they should be decreased about 10 percent†

Volume of Coarse Aggregate Per Cubic Yard of Concrete
Table 3-7

EXAMPLE OF THE ABSOLUTE VOLUME METHOD

The mix proportions are to be determined for the following conditions:

Maximum water-cement ratio	5.5 gallons per sack
Maximum size of aggregate	¾ inch
Air content	6 ± 1 percent
Slump	2 to 3 inches
Finenes modulus of fine aggregate	2.75
Specific gravity of portland cement	3.15
Specific gravity of fine aggregate	2.66
Specific gravity of coarse aggregate	2.61
Dry-rodded unit weight of coarse aggregate	104 pounds per cubic foot

The water requirement is estimated at 34 gallons per cubic yard from table 3–2. Since the maximum water-cement ratio is 5.5 gallons per sack, the cement factor must be at least 6.2 sacks per cubic yard. The volume of dry-rodded coarse aggregate is estimated to be 16.9 cubic feet per cubic yard from table 3–7. Thus the weight of coarse aggregate is 16.9 x 104 = 1,758 pounds. The absolute volumes of these quantities of materials is now calculated by use of the relationship:

$$\text{absolute volume} = \frac{\text{weight of material}}{\text{specific gravity} \times \text{unit weight of water}}$$

The weight of water is 62.4 pounds per cubic foot, and water weighs approximately 8.33 pounds per gallon. Thus the absolute volumes are—

Cement
$$\frac{6.2 \times 94}{3.15 \times 62.4} = 2.97 \text{ cubic feet}$$

Water
$$\frac{34 \times 8.33}{1 \times 62.4} = 4.54 \text{ cubic feet}$$

Coarse aggregate
$$\frac{1{,}758}{2.61 \times 62.4} = 10.80 \text{ cubic feet}$$

Air
$$.06 \times 27 = 1.62 \text{ cubic feet}$$
$$\text{Total} = 19.93 \text{ cubic feet}$$

The absolute volume of the fine aggregate is—

$$27.00 - 19.93 = 7.07 \text{ cubic feet}$$

and its weight is—

$$7.07 \times 2.66 \times 62.4 = 1{,}170 \text{ pounds.}$$

These estimated quantities should be used for the first trial batch. Subsequent batches should be adjusted by maintaining the water-cement ratio constant and achieving the desired slump and air content.

VARIATION IN MIXES

The proportions arrived at in determining mixes will vary somewhat depending upon which method is used. This occurs because of the empirical nature of these methods. It does not necessarily imply that one method is better than another. Each method begins by assuming certain needs or requirements and then proceeding to determine the other variables. Since the methods begin differently and use different procedures, the final proportions vary slightly. This is to be expected, and it further points out the necessity of trial mixes in determining the final mix proportions.

CHAPTER 4

EXCAVATION

SOIL CONDITIONS

The most important condition in the location of footings is the character of the soil on which they are to rest. Since the earth is made up of different types of strata or soils, consideration must be given to their suitability as a foundation base. Tests made every ten feet will show definitely the character of the soil upon which the foundation is to rest. The soil or strata generally encountered in small building foundations may be classified into three divisions: Rocks, Soil, and Fill.

1. Rock. Rock is the best foundation upon which to build, provided that it is made level. Otherwise, it must be stepped-up with concrete to a level footing. The rock should underlie the whole foundation.

2. Virgin Soil. Virgin soil is clay, gravel, loam, sand, or marshy ground in its natural condition. Clay is the most uncertain of soils because of its elasticity. In dry seasons, it is very firm while in wet seasons it becomes very slippery and unreliable. Clay is usually found in layers extending over a large area. Gravel, when compact and mixed with sharp sand to form a firm and unyielding stratum, is known as hard pan. This makes an excellent foundation. Loam, when mixed with other earthy substances, and when compact and of considerable depth, is a good foundation soil. Sand is formed by the decom-

position of rocks. It is generally found in beds which were, at one time, river beds. It is not a good foundation soil unless it is well confined and well drained. Quick-sand is a very fine sand mixed with loamy materials which retains a large quantity of water. It is very poor foundation soil as it will flow into the excavation as fast as it can be dug out. Quick-sand can be prepared for foundations only with great difficulty and expense. Marshy soils are formed by the decay of plants and other vegetable matter in sluggish water. Since there is no current, the plants take root in the bed. Successive beds of decayed plants are formed under slight pressure, and small cavities are formed. In some cases, these beds have become so deep that their bottoms have never been reached. This type of soil is sometimes called bog or swamp and is very dangerous foundation soil.

3. Fill. Made or artificial soil is often found in subdivisions or where depressions in the earth have been filled with refuse or city dumpings. It should not be built upon until the excavations is cleaned of the poor fill and the proper soils on which to rest the footings are reached.

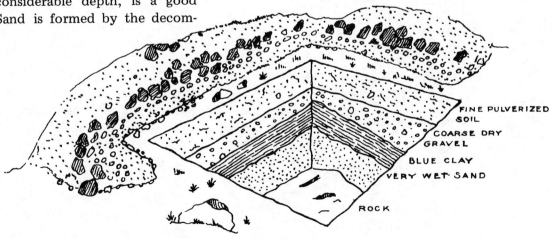

Pit Method of Determing Soil Conditions
Figure 4-1

PRELIMINARY CONSIDERATIONS

To evaluate the soil, dig a square pit about two feet larger than the proposed concrete structure. The pit should be deep enough to reach good foundation soil below the frost line of the locality, (Figure 4-1).

Observe the sides of the pit and note if there are any layers of soil that will shed surface and subsurface water into the pit. Such a layer is shown in Figure 4-2 where the footing rests on a layer of blue clay and the subsurface water flows into the footing bed.

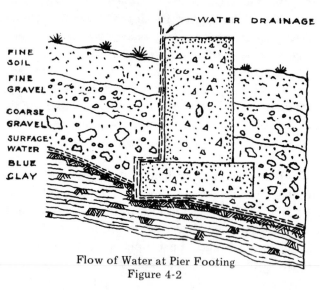

Flow of Water at Pier Footing
Figure 4-2

If there are soil layers present that will shed surface or subsurface water into the pit, the excavation should be drained as shown in Figure 4-3.

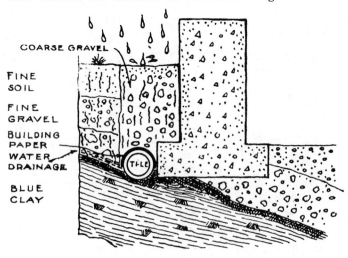

Drainage of Water
Figure 4-3

In Figure 4-3, a trench one foot wide has been allowed on the outside of the wall or pier. Place drain tiles at the bottom of this trench and slope the tiles to a point outside the foundation. Lay them with an opening of about one half inch at the joints. Embed the bottom of each joint in cement and cover the top of each joint with building paper. Fill the trench with coarse gravel to within about six inches of the grade line. Cover the gravel with top soil.

Surface water seeps into the ground until it strikes a layer of clay. It then flows along the top of the clay along the natural slope, forming a slippery surface. A footing placed on the clay is likely to move if enough surface and sub-surface water flows along the bed of clay. For this reason, soils surrounding foundations should be drained of excessive water. In the construction of housing foundations where many piers are needed for a single building, it is better to drain the outside of the whole foundation rather than to drain each pier separately. No matter how carefully a foundation is built, it is likely to shift if water is not prevented from accumulating in the foundation bed.

If a group of pier footings must be drained separately, dig a trench around the outside of the pier footings as shown in section in Figure 4-3 and in Figure 4-4.

Lay the tile as directed for wall footings, and slope the tile to a catch basin outside the foundations.

In building sites where the amount of water is excessive, drain tiles should be laid diagonally under the foundation. Care must be taken that the drain tiles do not run underneath a pier footing.

There are several methods of determining soil conditions. The open pit method consists of digging a pit at the approximate location of the proposed foundations and to a depth below the frost line where a good foundation soil is reached. By observing the side wall of the pit, the location and type of each layer of soil may be studied and classified as to its suitability as a foundation soil. This method is used to determine how deep the footing must be placed so it will rest on a good foundation soil. It will then be possible to determine the size of structure necessary to carry the proposed load.

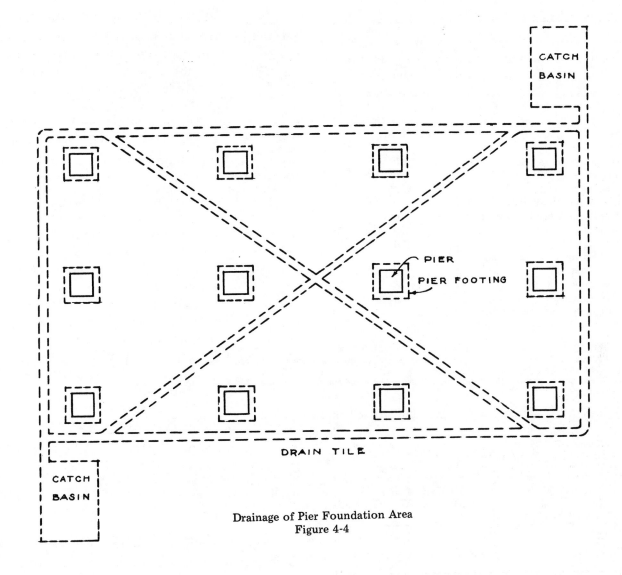

Drainage of Pier Foundation Area
Figure 4-4

SIZING THE FOUNDATION

The foundation should be capable of supporting about twice the design load of the planned structure without showing excessive settling. Local building codes often prescribe the allowable design bearing values for various types of foundation materials. The data in Figure 4-5 can be used for most applications.

A common cause of foundation movement is the failure to place the footings below the frost line. If the temperature drops to below freezing before the water drains from the soil, the frost penetrates the soil deeply, thus freezing the water in the soil to a considerable depth. The resulting expansion of the soil under frost pressure causes heaving. Water helps carry the frost line down into the earth. If the water drains from the soil rapidly, the frost cannot penetrate the ground very far. A good example of this is under a porch where the ground is dry. Even in zero weather, the top of the soil will not be frozen as long as it is dry. It will heave and harden under freezing conditions, according to the amount of moisture it contains. In most cities the building code assumes a depth of frost line for that community. Twenty to thirty inches below the surface is considered safe in most cases. If the working drawings do not show finished grades, depths of footings may be determined using Table 4-1.

Take care not to excavate below prescribed depths, but if this happens, place concrete to the depth actually excavated. Do not refill excavations to the specified depth before placing concrete. It is

33

CLASS OF MATERIAL	MINIMUM DEPTH OF FOOTING BELOW ADJACENT VIRGIN GROUND	VALUE PERMISSIBLE IF FOOTING IS AT MINIMUM DEPTH. POUNDS PER SQUARE FOOT	INCREASE IN VALUE FOR EACH FOOT OF DEPTH THAT FOOTING IS BELOW MINIMUM DEPTH. POUNDS PER SQUARE FOOT	MAXIMUM VALUE. POUNDS PER SQUARE FOOT
1	2	3	4	5
Rock	0' 0"	20% of ultimate crushing strength	0	20% of ultimate
Compact coarse sand	1' 0"	1500*	300*	8000
Compact fine sand	1' 0"	1000*	200*	8000
Loose sand	2' 0"	500*	100*	3000
Hard clay or sandy clay	1' 0"	4000	800	8000
Medium stiff clay or sandy clay	1' 0"	2000	200	6000
Soft sandy clay or clay	2' 0"	1000	50	2000
Adobe	1' 6"	1000**	50	
Compact inorganic sand and silt mixtures	1' 0"	1000	200	4000
Loose inorganic sand and silt mixtures	2' 0"	500	100	1000
Loose organic sand and silt mixtures and muck or bay mud	0' 0"	0	0	0

*These values are for footings one foot in width and may be increased in direct proportion to the width of the footing to a maximum of three times the designated value.
**For depths greater than eight feet (8') use values given for clay of comparable consistency.

Average allowable soil pressure (pounds per square foot)
Figure 4-5

TEMPERATURE ZONE (F.)	MINIMUM DEPTH BELOW EXISTING GRADE (FEET)
+20°	2
0	3
-20°	4

Minimum Excavation Depths
Table 4-1

too difficult to compact the fill surface properly. Since forms are not needed for footings in stable soils such as clay or disintegrated rock, footing excavations in these soils may be the actual width of footings as shown on drawings.

PREPARING BASE FOR SLABS ON GRADE

Bases for concrete slabs on grade consist of well compacted fill of crushed stone, sand, gravel, or cinders which have been wetted down and tamped to designed grade and line. Cinders when wet produce acids which are destructive to some types of utility lines; take adequate precautions to prevent such damage.

The base includes forming for utility ducts or drainage trenches, interior integral footings, and equipment or machinery pads as needed. Sewage lines, drain lines, and other utilities may be placed in the fill. Fill should be placed on stable undisturbed bearing soil. Subsequent to grading and prior to concrete placement the fill should be suitably treated to protect against insects and rodents.

FENCING AND SHORING

If not already protected, erect substantial railings near the edges of excavations to safeguard personnel and property. Shore excavation side walls against lateral movement and cave-ins, if the type of soil so requires (Figure 4-5). Inspect buildings and structures adjacent to new excavations to determine the need for shoring or other protective measures.

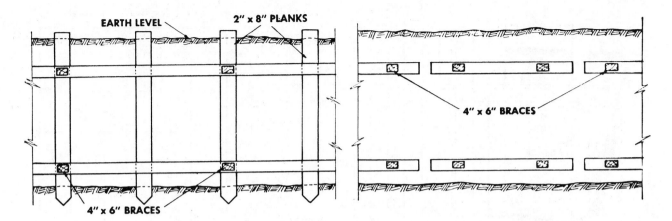

SKELETON STAY BRACING

HORIZONTAL STAY BRACING

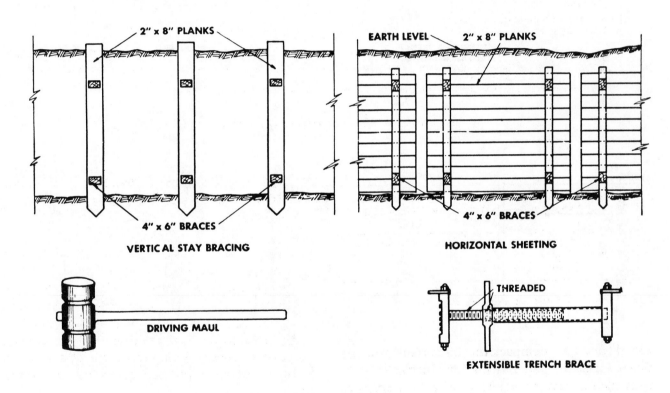

VERTICAL STAY BRACING

HORIZONTAL SHEETING

DRIVING MAUL

EXTENSIBLE TRENCH BRACE

STEEL DRIVING BLOCK

1 Trench-bracing methods

Bracing and shoring
Figure 4-5

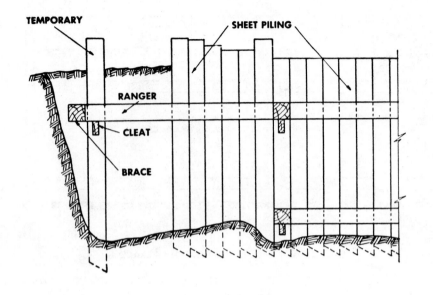

SECTION OF A TRENCH
SHEET-PILING METHOD

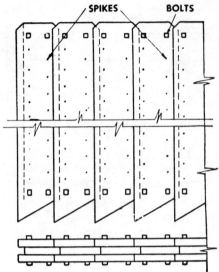

WAKEFIELD SHEET-PILING WOOD

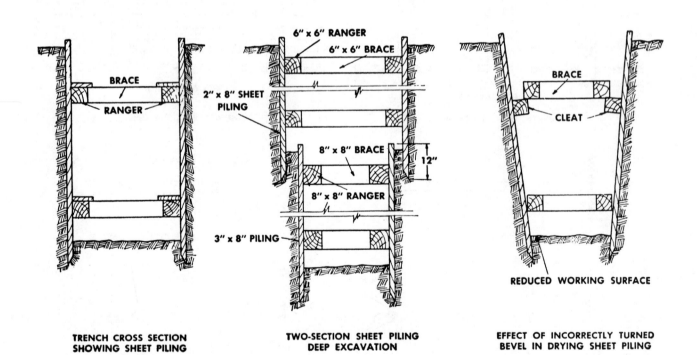

TRENCH CROSS SECTION
SHOWING SHEET PILING

TWO-SECTION SHEET PILING
DEEP EXCAVATION

EFFECT OF INCORRECTLY TURNED
BEVEL IN DRYING SHEET PILING

2 Trench-shoring methods

Bracing and shoring — continued
Figure 4-5

36

MAN HOURS REQUIRED FOR EXCAVATION

The rate of hand excavation is determined from Table 4-2. It varies with the type of soil and the depth of excavation. When mechanical equipment is used, the last six inches of bottom excavation must be cleared out and shaped by hand because it is extremely difficult to accurately excavate by machine.

Machine excavation is a neccesity for large projects which require substantial excavation. Types of excavation equipment which are particularly suitable for use in concrete construction work include power shovels, dragline buckets, and backhoes. Considerations which enter into the selection of equipment are the total yardage to be moved, working time available, type of excavation, and nature of the area. Due to the many variables, it is not possible to give generalized rates of excava-

Equipment	Type of materials	Average output cu yd/hr
Power shovel (½ cu yd capacity)	Sandy loam	70
	Common earth	60
	Hard clay	45
	Wet clay	25
Short-boom dragline (½ cu yd capacity)	Sandy loam	65
	Common earth	50
	Hard clay	40
	Wet clay	20
Backhoe (⅓ cu yd capacity)	Sandy loam	55
	Common earth	45
	Hard clay	35
	Wet clay	25

*90° swing, closed pit such as basement.

Earth Excavation by Machine*
Table 4-3

tion for various types of equipment. Some typical rates of excavation for specific conditions are given in Table 4-3. In practice there will be considerable variation from these rates.

Type of material	Cubic yards per man-hour					
	Excavation with pick and shovel to depth indicated				Loosening earth— man with pick	Loading in trucks or wagons— one man with shovel and loose soil
	0 to 3 foot	0 to 6 foot	0 to 8 foot	0 to 10 foot		
Sand	2.0	1.8	1.4	1.3	—	1.8
Silty sand	1.9	1.6	1.3	1.2	6.0	2.4
Gravel, loose	1.5	1.3	1.1	1.0	—	1.7
Sandy silt-clay	1.2	1.2	1.0	.9	4.0	2.0
Light clay	.9	.7	.6	.7	1.9	1.7
Dry clay	.6	.6	.5	.5	1.4	1.7
Wet clay	.5	.4	.4	.4	1.2	1.2
Hardpan	.4	.4	.4	.3	1.4	1.7

Earth Excavation by Hand
Table 4-2

CHAPTER 5

LAYING OUT THE BUILDING

Before building lines are established, the lot should be surveyed by a registered surveyor. Existing lot lines or stakes should not be depended on unless they have been checked by the surveyor.

The builder should find out from the building code of the community and the map of the section if there are any restrictions concerning the location of the building on the lot. A knowledge of the grade line established at that particular lot is also necessary to place the footings of the foundations at the proper levels.

Some builders use temporary stakes on which to stretch lines showing the outline of the excavation. This may answer the purpose for approximate work but it is better practice to erect batter boards carefully at the corners of the excavation. They should be placed far enough away from the excavation so as not to interfere with the digging. They should be well braced and the cross bars should afford reliable work points on which to stretch lines indicating the outside line of the excavation, and the outside and inside of the masonry wall or concrete forms. The tops of the batter boards are sometimes notched to show these points so that the lines may be taken down and replaced in the same locations. The boards may be leveled to coincide with some point on the finished wall, such as the top of the wall, or the first floor line of the building.

The easiest, quickest and most accurate way to determine the building lines of a new building is to use a builder's transit. When such an instrument is not available, square corners may be laid out by means of the 6—8—10 rule together with batter boards at each corner of the excavation. This rule is based on the fact that a triangle whose sides are exactly 6, 8 and 10 feet long is a right angle triangle, the angle opposite the longest side being a 90° angle (Figure 5-1).

If there is to be a cellar excavation and if the house sewer line is to be under the cellar floor, it

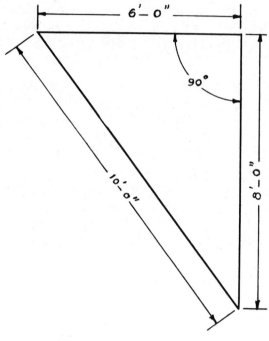

6-8-10 Rule
Figure 5-1

is well to know the elevation of the sewer line of the street. The cellar floor then can be placed at a higher level than the street sewer to afford proper sewer drainage. Grade elevations should be established at the time of laying out the foundations, and should be kept in reasonable harmony with adjacent property and street curb elevation. Otherwise, surface water drainage from one parcel of land to another might result.

LAYOUT PROCEDURE

It is of great importance in the layout of building lines that the work be checked and rechecked, and that the dimensions shown on the drawings prove themselves. This procedure can best be accomplished by erecting rigid batter boards and by checking all measurements from lot lines established by a registered surveyor. The following procedure should produce highly satisfactory results:

1. Establish a base line from the surveyor's lines to mark out one end of the building. See line AB, Figure 5-2. Set stakes locating the two corners at

A and B on this line.

2. Drive a nail in the center of the top of each of these stakes to locate the corners accurately.

3. On the line AB drive a stake F, 6 feet from stake A. Drive a nail in the top of this stake exactly 6 feet from the mail in stake A.

4. Drive stake E so a nail driven in the top will be exactly 8 feet from the nail in stake A and exactly 10 feet from the nail in stake F. The corner (EAF) will then be a right angle.

5. With a steel tape, lay out the length of the building along line AE extended to D which will be the third corner. Drive a stake at this point. A nail driven in the top of this stake should accurately mark the end of the boundary line AD.

6. Hook the end of a steel tape on the nail on stake D and lay off the width of the building toward C.

8. Erect batter boards at the four corners but far enough away so they will not be disturbed by the excavating. Set the tops of the batter boards at first floor level or at some other convenient point.

NOTE: In establishing the elevation of any part of the foundation, the top of the curb stone of the street may be used as a reference point. If there is no curb, the surveyor should establish that elevation to govern the location of the top of the batter boards. A carpenter's leveling stand is sometimes used to transfer the curb level to one batter board.

9. Stretch lines above the nails in stakes A, B, C and D, Figure 5-2 and fasten them to nails driven in the batter boards.

10. Test the squareness of the corners by one 6—8—10 method. Test the complete layout by measuring the diagonals with a steel tape (Figure 5-3). If the diagonals are equal in length, the corners are square.

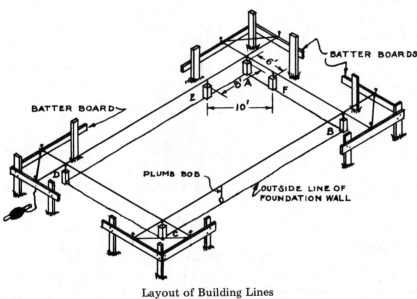

Layout of Building Lines
Figure 5-2

11. After the batter boards have been set in position and after the lines indicating the layout of the building have been transferred to them, remove the corner stakes (A, B, C and D, Figure 5-2.

7. Hook the end of another steel tape over the nail on stake B and measure the length of the building toward C. When both steel tapes are pulled taut and the points on the tapes showing the length and width of the building coincide, the point C will be accurately located. Drive a stake at this point and drive a nail in the top of the stake accurately locating this point.

12. To locate points from the suspended lines to the surface of the ground or to the bottom of an excavation, use a plumb-bob as shown in Figure 5-2. In this case the outside foundation line is located from the suspended line. In Figure 5-3 the corner stakes at the bottom of the excavation are located in the same manner.

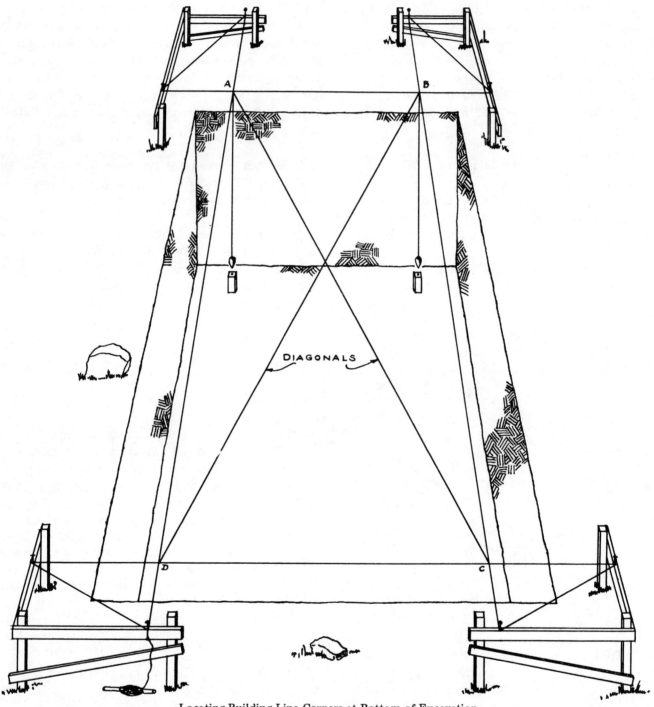

Locating Building Line Corners at Bottom of Excavation
Figure 5-3

40

CHAPTER 6

DESIGN OF CONCRETE FORMS

Formwork may represent as much as 33% of the total cost of a concrete structure, so the importance of the design and construction of this phase of a project cannot be overemphasized. The character of the structure, availability of equipment and form materials, anticipated repeated use of the forms, and familiarity with methods of construction influence design and planning of the formwork. Forms must be designed with a knowledge of the strength of the materials and the loads to be carried. The ultimate shape, dimensions, and surface finish must also be considered in the preliminary planning phase. Forms also provide protection for concrete, aid in curing, and support reinforcing rods and conduit which may be embedded within the concrete.

Forms for concrete structures must be tight and strong. If forms are not tight, there will be a loss of mortar which may result in honeycomb, or a loss or water that causes sand streaking. The forms must be braced enough to stay in alinement, and strong enough to hold the concrete. Special care should be taken in bracing and tying down forms, such as those for retaining walls, in which the mass of concrete is large at the bottom and tapers toward the top. In this type of construction and in other types, such as the first pour for walls and columns, the concrete tends to lift the form above its proper elevation. If the forms are to be used again they must be easily removed and re-erected without damage. Most forms are made of wood but steel forms are commonly used for work involving large unbroken surfaces, such as retaining walls, tunnels, pavements, curbs, and sidewalks. Steel forms for sidewalks, curbs, and pavements are especially advantageous since they can be used many times.

Forms are generally constructed from four different materials:

1) *Earth.* Earth forms are used in subsurface construction where the soil is stable enough to retain the desired shape of the concrete structure. The advantages of this type of form are that less excavation is required and there is better settling resistance. The obvious disadvantage is a rough surface finish, so the use of earth forms is generally restricted to footings and foundations.

2) *Metal.* Metal forms are used where added strength is required or where the construction will be duplicated at another location. Metal forms are more expensive, but they may be more economical than wooden forms if they can be used often enough. Examples of their use would be highway paving forms or curb and sidewalk forms.

3) *Wood.* Wooden forms are by far the most common type used in building construction. They have the advantage of economy, ease in handling, ease of production, and adaptability to many desired shapes. Added economy may result from reusing form lumber later for roofing, bracing, or similar purposes. Lumber should be straight, structurally sound, strong, and only partially seasoned. Kiln-dried timber has a tendency to swell when soaked with water from the concrete. If the boards are tight jointed the swelling causes bulging and distortion. If green lumber is used allowance should be made for shrinkage or the forms should be kept wet until the concrete is in place. Soft woods such as pine, fir, and spruce make the best and most economical form lumber since they are light, easy to work with, and available in almost every region. Lumber that comes in contact with the concrete should be surfaced on one side at least and on both edges. The surfaced side is turned toward the concrete. The edges of the lumber may be square, shiplap, or tongue and groove. The latter makes a more watertight joint and tends to prevent warping. Plywood can be used economically for wall and floor forms if it is made with waterproof glue and is identified for use in concrete forms. Plywood is more warp-resistant and can be reused more often than lumber. Plywood is made in thicknesses of $\frac{1}{4}$, $\frac{3}{8}$, $\frac{9}{16}$, $\frac{5}{8}$, and $\frac{3}{4}$ of an inch and in widths up to 48 inches. Although longer lengths are manufactured, 8-foot lengths are the most commonly used. The $\frac{5}{8}$-and $\frac{3}{4}$-inch thicknesses are most economical; the thinner sections will require solid backing to prevent deflection. The $\frac{1}{4}$-inch thickness is useful for curved surfaces.

4) *Fiber Forms.* Impregnated and waterproofed cardboard and other fiber materials are used as forms for round concrete columns and other applications where preformed shapes are desirable. These forms are usually made by gluing successive layers of fiber together and molding them to the desired shape. A major advantage of these forms is the time saved since form fabrication at the job site is not necessary.

Wall Forms. Elements of wooden forms for a concrete wall are shown in figure 6-1.

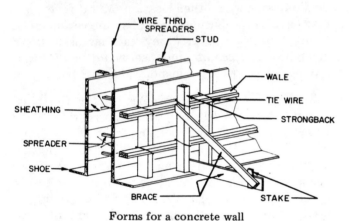

Forms for a concrete wall
Figure 6-1

(1) *Sheathing.* Sheathing forms the surfaces of the concrete. It should be as smooth as possible, especially if the finished surfaces are to be exposed. Since the concrete is in a plastic state when placed in the form, the sheathing should be watertight. Tongue and groove sheathing gives a smooth watertight surface. Plywood or hardboard can also be used.

(2) *Studs.* The weight of the plastic concrete will cause the sheathing to bulge if it is not reinforced. Studs are run vertically to add rigidity to the wall form. Studs are generally made from 2 x 4 or 3 x 6 material.

(3) *Wales (Walers).* Studs also require reinforcing when they extend over four or five feet. This reinforcing is supplied by *double wales.* Double wales also serve to tie prefabricated panels together and keep them in a straight line. They run horizontally and are lapped at the corners of the forms to add rigidity. Wales are usually made from the same material as the studs.

(4) *Braces.* There are many types of braces which can be used to give the forms stability. The most common type is a diagonal member and horizontal member nailed to a stake and to a stud or wale. The diagonal member should make a 30°

angle with the horizontal member. Additional bracing may be added to the form by placing vertical members (strongbacks) behind the wales or by placing vertical members in the corner formed by intersecting wales. Braces are not part of the form design and are not considered as providing any additional strength.

(5) *Shoe plates.* The shoe plate is nailed into the foundation or footing and is carefully placed to maintain the correct wall dimension and alinement. The studs are tied into the shoe and spaced according to the correct design.

(6) *Spreaders.* In order to maintain proper distance between forms, small pieces of wood are cut to the same length as the thickness of the wall and are placed between the forms. These are called spreaders. The spreaders are not nailed but are held in place by friction and must be removed before the concrete hardens. A wire should be securely attached to the spreaders so that they can be pulled out after the concrete has exerted enough pressure to the walls to allow them to be easily removed.

(7) *Tie wires.* Tie wire is a tensile unit designed to hold the concrete forms secure against the lateral pressure of unhardened concrete. A double strand of tie wire is always used.

Column Forms. Elements of wooden forms for concrete columns are shown in figure 6-2.

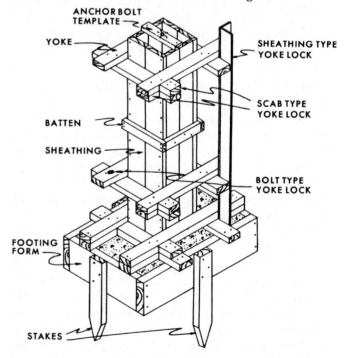

Form for a concrete column
Figure 6-2

(1) *Sheathing.* In column forms, sheathing runs vertically to save on the number of sawcuts required. The corner joints should be firmly nailed to insure watertightness.

(2) *Batten.* Batten are narrow strips of boards (cleats) that are placed directly over the joints to fasten the several pieces of vertical sheathing together.

(3) *Yokes.* The horizontal dimensions on a column are small enough so that bracing is not required in the vertical plane. A rectangular horizontal brace known as a yoke is used. The yoke wraps around the column and keeps the concrete from distorting the form. The yoke can be locked by the sheathing, scab, or bolt type yoke lock.

DESIGN CONSIDERATIONS

Forms for concrete construction must support plastic concrete until it has hardened. Stiffness is an important feature in forms and failure to provide for this may cause unfortunate results. Forms must be designed for all the weight they are liable to be subjected to including the dead load of the forms, plastic concrete in the forms, the weight of workmen, weight of equipment and materials whose weight may be transferred to the forms, and the impact due to vibration. These factors vary with each project but none should be neglected. Ease of erection and removal are also important factors in the economical design of forms. Platform and ramp structures independent of formwork are sometimes preferred to avoid displacement of forms due to loading and impact shock from workmen and equipment.

When concrete is placed in the form, it is in a plastic state and exerts hydrostatic pressure on the forms. The basis of form design, therefore, is the maximum pressure developed by the concrete during placing. The maximum pressure developed will depend on the rate of placing and the temperature. The rate of placing will affect the pressure because it determines how much hydrostatic head will be built up in the form. The hydrostatic head will continue to increase until the concrete takes its initial set, usually in about 90 minutes. However, at low temperatures, the initial set takes place much more slowly so it is necessary to consider the temperature, at the time of placing. Knowing these two factors, and the type of form material to be used, a tentative design may be calculated.

When the forms are to be used for only one foundation, and the form material is to be re-used in other parts of the building, the studs and walers are not cut to length but are left in stock lengths. The sheathing is nailed only enough to hold it in place. It can then be easily taken apart. After the walls have been stripped, the lumber should be cleaned of cement and piled in such a way as to prevent it from warping.

Forms are sometimes made up in standard size panels. This method is used where there are many foundations of a similiar size to be built. The panels are made in lengths of from 2 to 12 feet and in widths of from 4 to 6 feet depending on the size of the wall. The use of these panels saves time in erecting the forms.

It will require thought and ingenuity on the part of the carpenter to plan efficient and economical forms. They should be made in sections that can be handled and set without difficulty and that can be fastened at the joints so they may be taken apart without breaking.

DESIGN OF WALL FORMS

Procedure. Is it desirable to design forms according to a step-by-step procedure to assure the consideration of all pertinent factors. Wooden forms for a concrete wall should be designed by the following steps:

(1) Determine the materials available for sheathing, studs, wales, braces, shoe plates and tie wires.

(2) Determine the rate (vertical feet per hour) of placing the concrete in the form.

(3) Make a reasonable estimate of the placing temperature of the concrete.

(4) Determine the maximum concrete pressure by entering the bottom of figure 6-3 with the rate of placing. Draw a line vertically up until it intersects the correct concrete temperature curve. Read horizontally across from the point of intersection to the left side of the graph and determine the maximum concrete pressure.

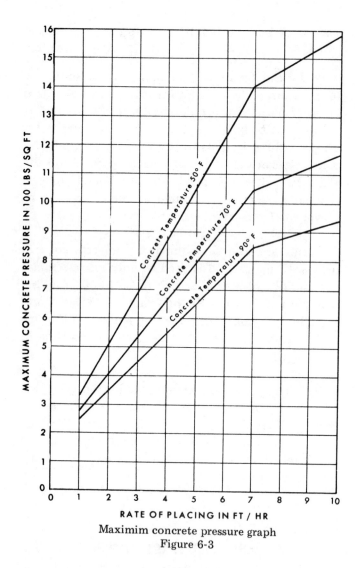

RATE OF PLACING IN FT / HR

Maximim concrete pressure graph
Figure 6-3

(5) Determine the maximum stud spacing by entering the bottom of figure 6-4 with the maximum concrete pressure. Draw a line vertically up until it intersects the correct sheathing curve. Read horizontally across from the point of intersection to the left side of the graph. If the stud spacing is not an even number of inches, round the value of the stud spacing down to the next lower even number of inches. For example, a stud spacing of 17.5 inches would be rounded down to 16 inches.

(6) Determine uniform load on a stud by multiplying the maximum concrete pressure by the stud spacing.

Uniform load on stud (lb/lineal ft) =
Maximum concrete pressure (lb/sq ft) ×
stud spacing (ft)

(7) Determine the maximum wale spacing by entering the bottom of figure 6-5 with the uniform load on a stud. Draw a line vertically up until it intersects the correct stud size curve. Read horizontally across from the point of intersection to the left side of the graph. If the wale spacing is not an even number of inches, round the value of the wale spacing down to the next lower even number of inches. Double wales (two similar members) are used in every case as shown in figure 6-1.

(8) Determine the uniform load on a wale by multiplying the maximum concrete pressure by the wale spacing.

Uniform load on wale (lb/lineal ft) =
Maximum concrete pressure (lb/sq ft) × wale spacing (ft)

(9) Determine the tie wire spacing, based on the wale size, by entering the bottom of figure 6-6 with the uniform load on a wale. Draw a line vertically up until it intersects the correct double wale size curve. Read horizontally across from the point of intersection to the left side of the graph. If the tie spacing is not an even number of inches, round the value of the tie spacing down to the next lower even number of inches.

(10) Determine the tie wire spacing based on the tie wire strength by dividing the tie wire strength by the uniform load on a wale. If the tie wire spacing is not an even number of inches, round the computed value of the tie spacing down to the next lower even numer of inches. If possible, use a tie wire size that will provide a tie spacing equal to or greater than the stud spacing. Always use a double strand of wire. If the strength of the available tie wire is unknown, the minimum breaking load for a double strand of wire is given in table 6-1.

Tie wire spacing (in) =
Tie wire strength (lbs) × (12 in/ft)

Uniform load on wale (lb/ft)	
STEEL WIRE	
Size of wire Gage No.	Minimum breaking load double strand Pounds
8	1700
9	1420
10	1170
11	930

Breaking Load of Wire
Table 6-1

(11) Determine the maximum tie spacing by selecting the smaller of the tie spacings based on the wale size and on the tie wire strength.

(12) Compare the maximum tie spacing with the maximum stud spacing. If the maximum tie spacing is less than the maximum stud spacing, reduce maximum stud spacing to equal the maximum tie spacing and tie at the intersections of the studs and wales. If the maximum tie spacing is greater than the maximum stud spacing, tie at the intersections of the studs and wales.

(13) Determine the number of studs for one

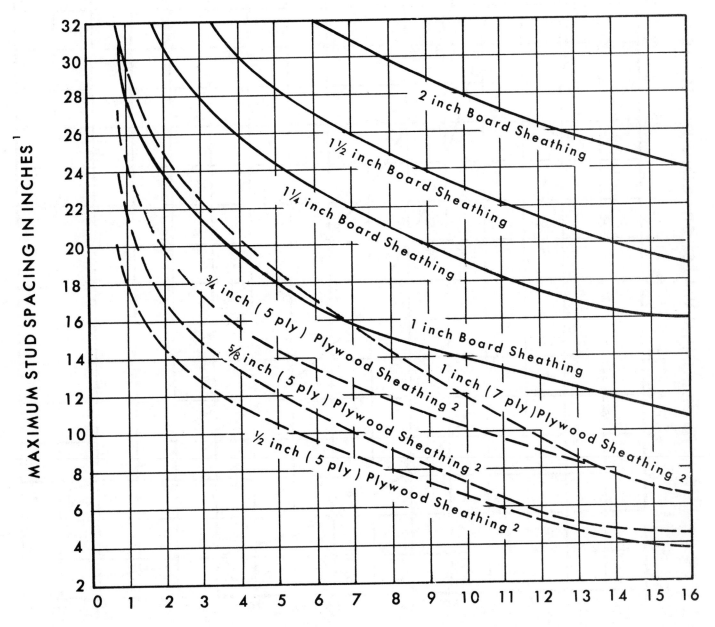

MAXIMUM CONCRETE PRESSURE IN 100 LBS/SQ FT

[1] MAXIMUM ALLOWABLE STUD SPACING = 32 INCHES

[2] SANDED FACE GRAIN PARALLEL TO SPAN

Maximum stud spacing graph
Figure 6-4

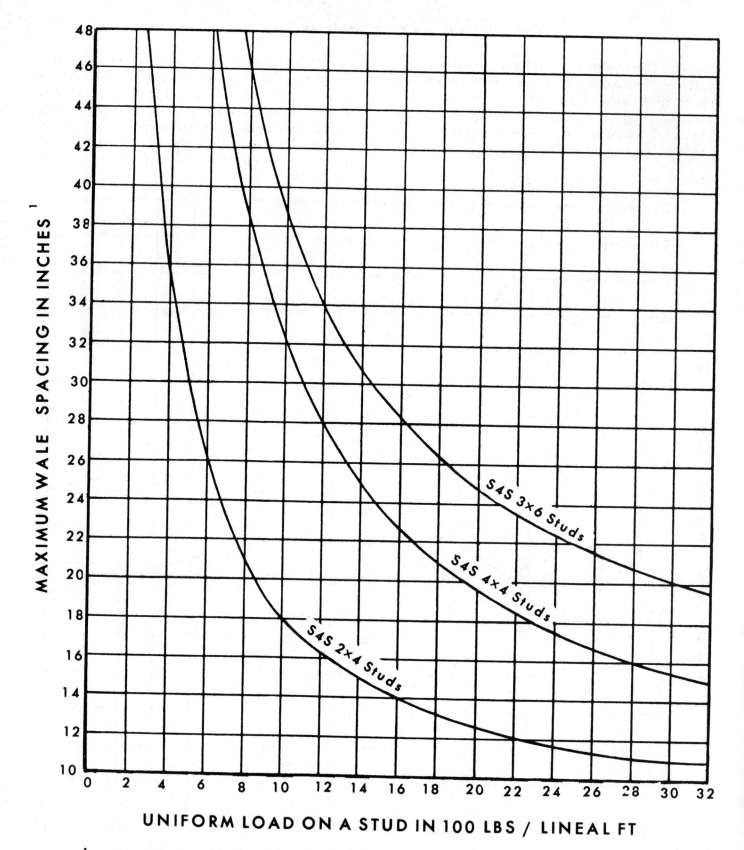

Maximum wale spacing graph
Figure 6-5

46

side of a form by dividing the form length by the stud spacing. Add one (1) to this number and round up the next integer. During form construction, place studs at the spacing determined above. The spacing between the last two studs may be less than the maximum allowable spacing.

$$\text{No. of studs} = \frac{\text{Legth of form (ft)} \times 12 \text{ (in/ft)}}{\text{Stud spacing (in)}} + 1$$

(14) Determine the number of wales for one side of the form by dividing the form height by the wale spacing and round up to the next integer. Place first wale one-half space up from the bottom and the remainder at the maximum wale spacing.

(15) Determine the time required to place the concrete by dividing the height of the form by the rate of placing.

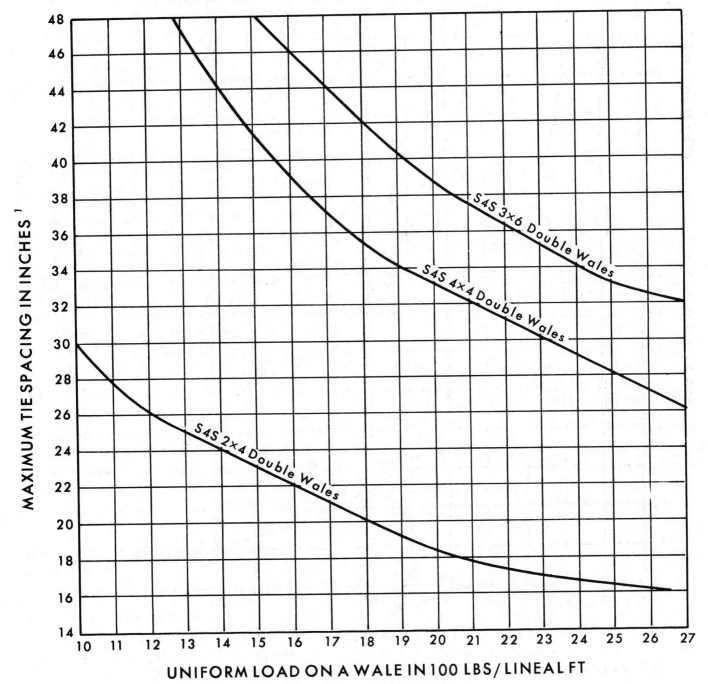

Maximum tie wire spacing
Figure 6-6

SAMPLE WALL FORM DESIGN PROBLEM

Design forms for a concrete wall 40 feet long, 2 feet thick, and 10 feet high. Concrete will be delivered and placed at the rate of 192 cubic feet per hour. The concrete temperature is estimated to be 70° F. The material to be used for forms is 2 x 4's and 1 inch board sheathing.

Solution Steps:

1. Material available: 2 x 4's, one inch sheathing and No. 9 wire.

2. Concrete placing rate: 192 cu. ft. /hr.

3. Plan area of forms = 40 ft × 2 ft = 80 sq ft

4. Rate of placing = $\frac{192 \text{ cu ft/hr}}{80 \text{ sq ft}}$ = 2.4 ft/hr

5. Temperature of concrete: 70° F.

6. Maximum concrete pressure (fig. 6-3) = 460 lb/sq ft

7. Maximum stud spacing (fig. 6-4) = 18 +, use 18 inches

8. Uniform load on studs = 460 lb/sq ft × $\frac{18 \text{ in.}}{12 \text{ in/ft}}$ = 690 lb/ft

9. Maximum wale spacing (figure 6-5) = 23 + use 22 inches

10. Uniform load on wales = 460 lb/sq ft × $\frac{22 \text{ in}}{12 \text{ in/ft}}$ = 843 lb/ft

11. Tie wire spacing based on wale size (fig. 6-6) = > 30 inches

12. Tie wire spacing based on wire strength = $\frac{1420 \text{ lb} \times 12 \text{ in/ft}}{843 \text{ lb/ft}}$ = 20 +—use 20 inches

13. Maximum tie spacing = 20 inches

14. Maximum tie spacing is greater than maximum stud spacing, therefore, reduce the tie spacing to 18 inches and tie at the intersection of each stud and double wale

15. Number of studs per side = (40 ft × $\frac{12 \text{ in/ft}}{18 \text{ in}}$) + 26.7 + 1, use 28 studs

16. Number of double wales per side = 10 ft × $\frac{12 \text{ in/ft}}{22 \text{ in}}$ = 5+—use 6 double wales

17. Time required to place concrete = $\frac{10 \text{ ft}}{2.4 \text{ ft/hr}}$ = 4.17 hrs.

COLUMN FORMS

Procedure. Wooden forms for a concrete column should be designed by the following steps:

(1) Determine the materials available for sheathing, yokes, and battens. Standard materials for column forms are 2 x 4's and 1-inch sheathing.

(2) Determine the height of the column.

(3) Determine the largest cross-sectional dimension of column.

(4) Determine the yoke spacings by entering table 6-2 and reading down the first column until the correct height of column is reached. Then read horizontally across the page to the column headed by the largest cross-sectional dimension. The center-to-center spacing of the second yoke above the base yoke will be equal to the value in the lowest interval that is partly contained in the column height line. All subsequent yoke spacings may be obtained by reading up this column to the top. This procedure gives maximum yoke spacings. Yokes may be placed closer together if desired. Table 6-2 is based upon use of 2 x 4's and 1-inch sheathing.

SAMPLE COLUMN FORM DESIGN PROBLEM

Example Problem: Determine the yoke spacing for a 9-foot column whose largest cross-sectional dimension is 36 inches. 2 x 4's and 1-inch sheathing are available.

Solution steps:

1. Material available—2 x 4's and 1-inch sheathing

2. Height of column is 9 feet

3. Largest cross-sectional dimension of the column is 36 inches.

4. Maximum yoke spacing for column (table 6-2) starting from the bottom of form are 8", 8", 10", 11", 12", 15", 17", 17" and 10". The space between the top two yokes has been reduced because of the limits of the column height.

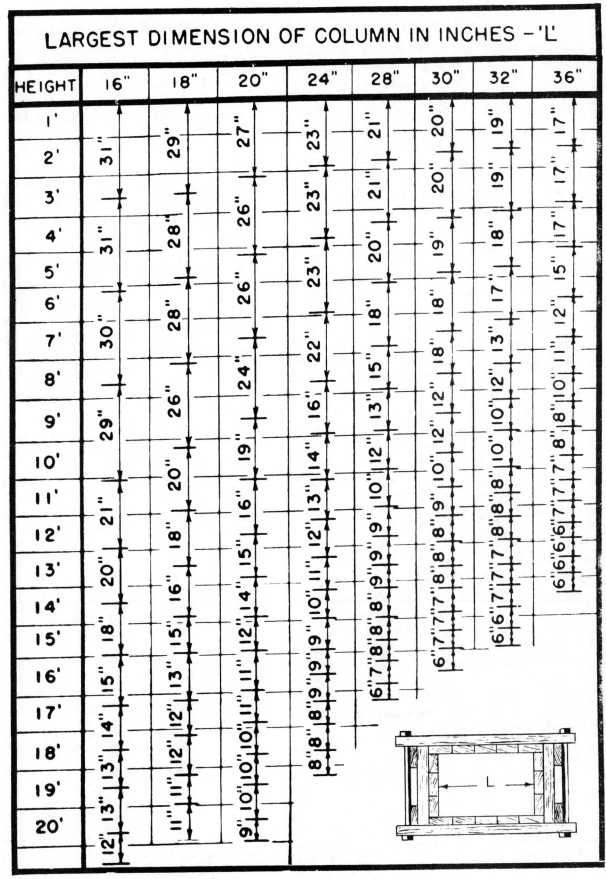

Column Yoke Spacing
Table 6-2

49

FOUNDATION FORMS

Foundation forms include forms for large footings, wall footings, column footings, and pier footings. These foundations or footings are relatively low in height and have a primary function of supporting a structure. The depth of concrete is usually small, therefore the pressure on the form is relatively low. Thus, design based on strength consideration generally is not necessary. Whenever possible, the earth should be evacuated and used as a mold for the concrete footings. Details for footing forms are given on page 68.

FLOOR FORMS

Procedure. Wooden forms for flat concrete slabs should be designed by the following steps:

(1) Determine material available for sheathing, cleats, joists, stringers, and shores. Typical materials are—1-inch tongue and groove or ¾-inch plywood for sheathing; 1- by 4-inch cleats, 2- by 6-inch joists; 2- by 8-inch, 4- by 4-inch, or 4- by 6-inch stringers; and 4- by 4-inch shores.

(2) Determine the total unit load on the floor form. The weight of ordinary concrete is assumed to be about 150 lb per cubic foot. Using this figure, the weight of concrete is 50 pounds per square foot for a 4-inch slab, 63 pounds per square foot for a 5-inch slab, and 75 pounds per square foot for a 6-inch slab. In addition, a live load for men and construction materials must be added. This is generally 50 pounds per square foot, however, 75 pounds per square foot is frequently used if powered concrete buggies are utilized.

(3) Determine the spacing of floor joists. Table 6-3 indicates the joist spacing as a function of slab thickness and span of the joists for 2- by 4-inch joists. This table gives spans determined by considering joist strength only. It does not take into consideration the deflection of the sheathing. If this is of concern, it should be checked by separate calculation. The joist span may be shortened by the addition of stringers.

Concrete slab thickness	Joist span						
	4-foot	5-foot	6-foot	7-foot	8-foot	9-foot	10-foot
4″ ----------	4′0″	4′0″	4′0″	4′0″	4′0″	3′0″	2′6″
5″ ----------	4′0″	4′0″	4′0″	3′6″	3′0″	3′0″	2′6″
6″ ----------	4′0″	3′0″	2′6″	2′0″	2′0″	2′0″	2′0″

*Based only on joist strength, does not consider deflection of sheathing.

Joint Spacing for 2- by 6-Inch Joist*
Table 6-3

(4) Determine the location of the stringers which support the joists. For short spans, it may not be necessary to use stringers.

(5) Determine the spacing of the shores, or posts, which support the stringers. Maximum spans for stringers are given in table 6-4.

Concrete slab thickness	Spacing of Stringers*					
	2 x 8″ stringer			4 x 6″ Stringer		
	5′	6′	7′	5′	6′	7′
4″	5′6″	5′0″	4′6″	6′6″	6′0″	5′6″
5″	5′6″	5′0″	4′6″	6′0″	5′6″	5′0″
7″	----	----	----	5′6″	5′0″	4′6″
8″	----	----	----	5′0″	4′6″	4′6″

*Spacing based on live load of 50 lb. per sq. ft. For a live load of 75 lb. per sq. ft., increase slab thickness by 2″ and use the corresponding spacing.

Maximun Spans for Stringers
Table 6-4

SAMPLE FLOOR FORM DESIGN PROBLEM

Example Problem. A 5-inch concrete floor slab has a span of 12 feet. Material available includes 1-inch tongue and groove sheathing, 1 x 4's, 2 x 6's, and 4 x 4's, and 2 x 8's. Determine the spacing of the joists, stringers, and shores.

Solution Steps:

1. Material available: 1-inch tongue and groove, 1 x 4's, 2 x 6's, 4 x 4's, and 2 x 8's.

2. Total unit load = live load + concrete load = 50 + 63 = 113 lb per sq ft

3. Spacing of joists (use 2 x 6's, table 6-3): Locate a stringer in the middle of the span, giving a joist span of 6 feet and a joist spacing of 4 feet.

4. Stringer spacing (use 2 x 8's) = 6 feet (from above).

5. Spacing of shores (use 4 x 4's, table 6-4) = 5 feet

STAIR FORMS

Various types of stair forms are used, including prefabricated forms. For moderate width stairs joining typical floors, design based on strength considerations is not generally necessary.

CHAPTER 7

FORM MATERIALS AND HOW TO USE THEM

LUMBER AND PLYWOOD

Various kinds of wood are used in building forms but the kind used in a particular case usually depends on the lumber dealer's stock on hand. However, softwood is most commonly used. In the East, fir and hemlock are generally used for studding and walers, while in the South and West very little hemlock is used.

ESTIMATING LUMBER REQUIREMENTS

Studs. To find the number of studs for a form in which the stud spacing is to be 24 inches O.C. (on centers), divide the length of the form by two and add one stud for the corner. For example, a form 100 feet long would need 51 studs (100 ÷ 2 = 50, 50 + 1 = 51). If the spacing is to be 16 inches O.C., divide by 1¼ instead of 2.

If 8 foot studs are to be used, the total length of the studs would be 8 x 51 = 408 feet. To this figure, add the lengths of the plates and walers to find the total number of lineal feet of 2 x 4's required. If the walers and plates for this form are 200 feet long, the total length of all of the 2 x 4's would be 608 feet.

Lumber is sold by board measure (B.M.). This is based on a unit 1 inch thick, 12 inches wide and 12 inches long. A piece 2 inches thick, 6 inches wide and 12 inches long would also be a board foot. A 2 x 4 which is 1 foot long contains 8/12 or 2/3 of a board foot. In the above example, the board measure would be 608 x 2/3, or approximately 405.

Sheathing. To find the amount of tongue and groove sheathing required, it is necessary to determine the number of square feet of surface to be covered. To this is added a certain percentage of this area to allow for matching and dressing. For 6 inch sheathing boards, this allowance to be added is 1/6 of the area. For example, approximately 1167 B.M. of tongue and groove sheathing would be needed to cover 1000 square feet. If plywood is to be used, no additions need be made to the actual area to be covered since 100 square feet of plywood will cover 100 square feet of surface.

Table 7-1 gives the coverage figures for common sheathing used for forms.

SQUARE EDGE BOARDS

NOMINAL SIZE	ACTUAL WIDTH DRESS	FACE	BOARD FEET REQUIRED TO COVER 100 S.F.*
1'' x 4''	3-1/2	3-1/2	114
1'' x 6''	5-1/2	5-1/2	109
1'' x 8''	7-1/4	7-1/4	110
1'' x 10''	9-1/4	9-1/4	108
1'' x 12''	11-1/4	11-1/4	107

TONGUE AND GROOVE

1'' x 4''	3-7/16	3-3/16	126
1'' x 6''	5-7/16	5-3/16	116
1'' x 8''	7-1/8	6-7/8	116
1'' x 10''	9-1/8	8-7/8	113
1'' x 12''	11-1/8	10-7/8	110

SHIPLAP

1'' x 6''	5-7/16	4-15/16	122
1'' x 8''	7-1/8	6-5/8	121
1'' x 10''	9-1/8	8-5/8	116
1'' x 12''	11-1/8	10-5/8	113

*Add 5% for waste

Coverage for sheathing
Table 7-1

Combination wall ties and spreaders are manufactured in accordance with the standard sizes of studding, walers and sheathing. Therefore, when stock lumber is used, the wall ties will be the correct length when in place.

Plywood Sheathing. Plywood panels may be secured in thicknesses of 1/8, 1/4, 3/8, 1/2, 5/8, and ¾ inches. These thicknesses are generally made up of three or five plies. These plies, in the type of plywood used for form construction, are held together by waterproof glue. The panels are 4 feet wide and range from 4 to 12 feet in length.

Masonite is another comparatively thin smooth material used for form covering. Both Masonite and plywood should be backed up as shown in Figure 7-1. Masonite, being flexible, requires more nailing than plywood.

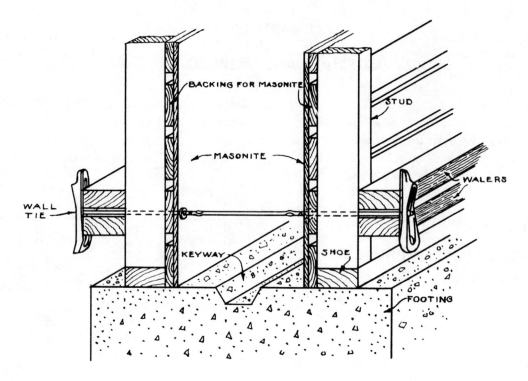

Wooden Form on Footing
Figure 7-1

Special plywoods made to resist moisture are used to some extent where the finished wall is to be free from fins or other irregularities. Because of its superior strength, its waterproof qualities, and the low cost of application, this special plywood is becoming popular for form construction.

LOCATING WALL FORMS ON A FOOTING

1. Establish a center line along the footing as shown by the symbol ₵ in Figure 7-2. A chalk line is generally used to mark this line.

2. Measure half the thickness of the proposed wall, from the center line on the footing toward the outside of the footing. This point indicates the outside of the wall or the inside of the sheathing as shown at A, Figure 7-2.

3. To find the line that indicates the inside face of the shoe, add the thickness of the sheathing to half the thickness of the wall, and measure this distance from the center line of the footing, as shown at B, Figure 7-2.

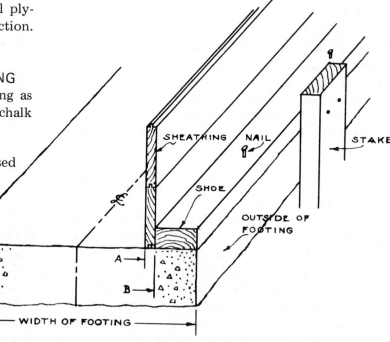

Locating Wall Forms on a Footing
Figure 7-2

NOTE: Use the above procedure to locate the outside form on the footing. To locate the inside form, proceed in the same manner, but measure in the opposite direction from the center line.

LOCATING FOOTING SHOES AND STUDS

Shoes.

NOTE: Material used for shoes in form work may be either 1 x 6 or 2 x 4.

1. Select straight stock if possible.

2. Fasten the shoe to the concrete by nailing, or hold it in position by bracing it from stakes driven in the ground along the edge of the footing (Figure 7-2).

How To Layout Studding. Layout stud spacing on the shoe. Spacing may range from 12 to 24 inches on centers, depending on the thickness of the wall.

How To Erect Studs.

1. Select studs that are free from defects and of the required length, making sure that they are straight and uniform in size.

2. Set all studs plumb and brace them temporarily. Braces should be 2 x 4 or 2 x 6 stock and should run at about a 30 degree angle from the horizontal.

HOW TO PLACE SHEATHING

Board Sheathing.

1. Fasten the sheathing to the framework of the form with common nails. Drive the nails at an angle of 90 degrees to the surface of the sheathing.

2. Nail the sheathing only in places where it is necessary to draw the joints tight and to secure the sheathing to the framework.

3. Nail each sheathing board at the bottom edge only, and nail each successive sheathing board at each alternate stud as shown in Figure 7-3.

4. Double nail the first sheathing board laid along the top of the footing at the shoe line. See D, Figure 7-3.

5. In case a joint occurs in the sheathing as at A, face nail the board B at its top edge and the board C at its bottom edge.

Plywood Sheathing.

1. Place the lower edge of all panels level, and place all joints on studs. Plywood less than 5/8 inch in thickness should be backed up with sheathing or stringers for support between the studs. Sheathing used for backing may be laid tight or the stringers may be spaced from 2 to 4 inches apart.

2. Nail the edges of plywood panels with flat headed nails spaced 6 to 12 inches. Space the nails about 18 inches in the rest of the panel.

NOTE: One quarter inch and three eighths inch plywood requires more nails to secure it than does tongue and groove sheathing. However, it should be nailed only enough to hold it in place.

HOW TO PLACE WALERS

NOTE: Place walers along the studs of long sections of a form or where there is a tendency for the form to bulge. The walers are used to align and strengthen the form and to provide a good bearing for the top of the side braces (Figure 7-4).

1. Place the sleepers in place and, if necessary, hold the outer end of each sleeper with a stake driven in the ground (Figure 7-4).

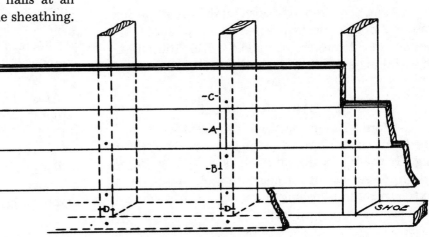

Face Nailing Sheathing
Figure 7-3

2. Cut the braces to approximate length.

3. Place the walers at the top of the braces.

4. Nail blocks of wood on the studs at the top of the walers.

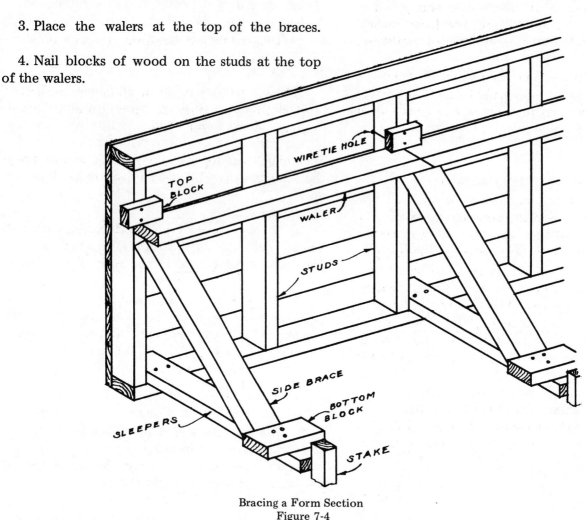

Bracing a Form Section
Figure 7-4

5. Plumb the form and nail a bottom block on the sleeper to hold the brace.

Nails. For concrete formwork where the forms are not to be nailed permanently, about 15 pounds of 8d nails are required for each 1000 B.M. of sheathing and studs.

GENERAL PRINCIPLES

The forms should be tight to prevent leakage of the liquid concrete through the joints of the sheathing. If the forms are not tight, ridges will be formed on the surface of the concrete wall, thus making extra work in finishing the wall. Ridges on the wall also make stripping of the forms difficult.

The carpenter should examine the forms and

their supports to detect any defects or weakness before the concrete is poured into the forms. There should be a constant check made on the braces while the concrete is being poured because, as the forms are filled, the pressure within them increases.

After the wooden forms have been built to the required measurements, they should be oiled to provide a smooth surface on the face of the concrete wall and to prevent the moist concrete from soaking into the pores of the wooden forms, causing the forms to stick to the hardened concrete wall and making the stripping more difficult.

After the forms have been set in the proper place, they should be fastened together, plumbed and braced on the corners, squared, and then aligned. In setting the forms, temporary braces should be put in place to hold the form straight until walers and braces can be properly secured.

After the forms have been removed from the hardened concrete, they should be carefully repaired where needed. If they are not to be set up at once, they should be stored for future use.

SPREADERS AND FASTENERS

Various types of combination metal wall ties and spreaders are used in form construction. The new type of rods with spreader washers set for the width of the wall, eliminates the necessity of using wood spreaders and saves considerable time in form building. The tie rods are drawn tight by means of metal wedges which can be tightened or adjusted from the outside of the form. Figure 7-5 shows a practical and popular type of rod and and spreader. The wedge tightener (Figure 7-6) is used with this type of tie rod.

How To Use The Hair Pin Type Of Tie Rod.

1. To use the tie rod shown in Figure 7-5, bore holes in the sheathing large enough to allow the eye end of the tie rod to pass through each side of the form.

2. Place the rods in the forms as each side of the form is erected.

NOTE: When placing sheathing on the second side of the form, bore the holes and insert the tie rods before placing the walers.

3. Place the walers as shown in Figure 7-5 so that the tie rod may enter between them.

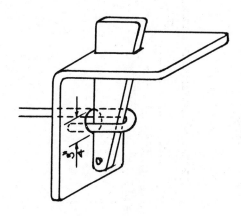

Enlarged View of Wedge Tightener
Figure 7-6

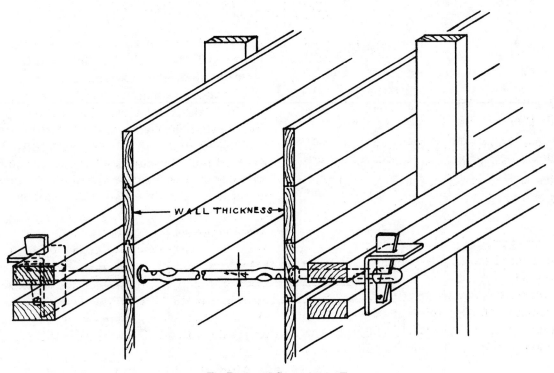

Tie Rod and Spreader in Form
Figure 7-5

4. Insert the wedge into the loop end of the rod and drive the wedge down until the spreader washers are tight against the inside of the sheathing.

Another type of spreader and tie rod is shown in Figure 7-7. The tightening wedge for this rod is shown at Figure 7-8 and the complete rod in position is shown in Figure 7-1.

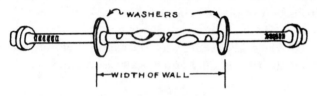

Tie Rod and Spreader
Figure 7-7

Wedge Tightener
Figure 7-8

The screw anchor type of wall tie (Figure 7-9) is used by some builders for heavy work. The tie or spreader is made of heavy wire, twisted and welded at each end (B) in such a way as to form threads into which the lag screws (A) may enter. The tie rods come in different lengths for walls of different thicknesses. Holes for the lag screws are bored through the sheathing between the studs. A lag screw is then placed through the walers and

into the threaded wall tie at B. The twisted ends (B), being larger than the hole through the sheathing, will not go through the sheathing. Thus, this device acts as a spreader and also as a tie.

How To Use The Lag Screw Type Of Tie Rod.

NOTE: Place the lag screw type of spreader (Figure 7-9) in the form after one side has been sheathed and the walers have been put in place.

1. Bore a hole in the sheathing and waler 1/16 inch larger than the lag screw.

2. Place the lag screw and a large washer in position thru the waler. Turn the wall tie on the lag screw.

3. Tie up the opposite side in the same manner after the sheathing and walers have been placed.

4. Tighten the lag screws with a wrench until the spreader is drawn tightly against the form siding.

NOTE: Keep the lag screw holes on opposite sides of the form at the same height; also keep the spreader at a right angle to the inside of the form.

Another type of tie rod used on heavy forms is shown in Figure 7-10A. A tie rod of this type may be made up on the job. This is a distinct advantage where there are many different sizes of walls. The 1/4 inch rod is cut to length on the job. The rods are drawn tight by means of a tightener designed for this type of wall tie (Figure 7-10B). The ends are held tight by means of a cast iron button provided with a locking screw which is tightened to hold the tension drawn with the tightener.

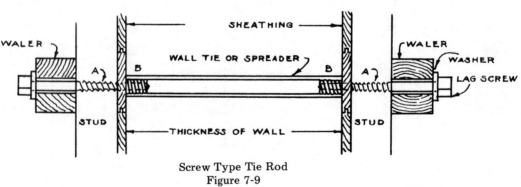

Screw Type Tie Rod
Figure 7-9

56

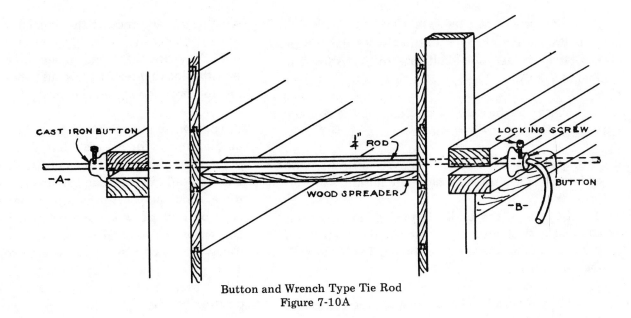

Button and Wrench Type Tie Rod
Figure 7-10A

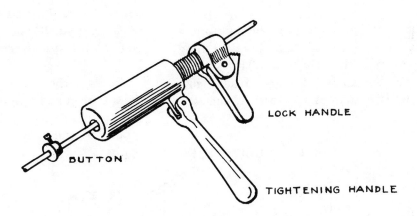

Rod Tightening Wrench
Figure 7-10B

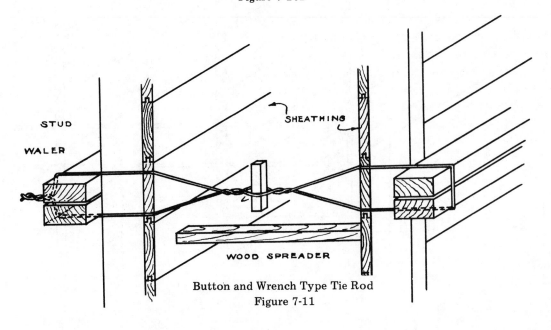

Button and Wrench Type Tie Rod
Figure 7-11

How To Use The Button Type Rod.

1. Bore holes through both sides of the sheathing large enough to admit the rod, as shown in Figure 7-10A.

2. Place the button on one end of the rod and tighten the set screw as shown at A, Figure 7-10A.

3. Slide the rod through the form until the tighted button is against the waler on one side.

4. Slide a button on the opposite end of the rod and place the tightening wrench (Figure 7-10B) on the rod behind and close to the locking button (B, Figure 7-10A).

5. Place a wood spreader between the forms and opposite the studs.

6. Hold the locking handle of the tightening wrench in the left hand and turn the tightening handle clockwise with the right hand. This will expand the wrench and draw the two buttons tight, thus pulling the two sides of the form against the spreader.

7. Tighten the set screw on the button at B, Figure 7-10A, and remove the wrench.

8. Bend down the end of the rod extending outside the locking button, because these ragged ends are dangerous to the workman.

The twisted tie wire is the oldest method of holding forms for concrete but, due to the amount of labor required to place and tighten them, they are used only in light and shallow forms. However, there are places on forms where the manufactured type of wall tie will not work, such as in angular corners where the tie is placed at an angle to the waler. In these cases, wire, when properly placed, is the most practical method to use. In Figure 7-11 a section of a form is shown with the wire in position as a tie.

How To Use Twisted Wire Ties.

1. Bore small holes through the sheathing at each side of the stud where the wire tie is to be placed.

2. Place the wire through the holes on both sides of the form and around the walers.

3. Pull the wires tight and twist the ends together.

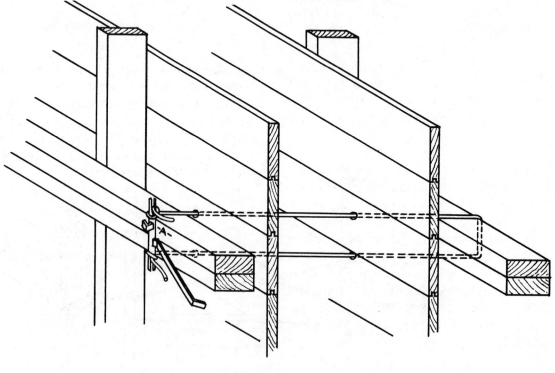

Wire Clamp
Figure 7-12

58

4.Place wood spreaders between the sides of the form and twist the wires by means of a stick until form walls are tight against the spreaders. (Figure 7-11).

Figure 7-12 shows another method of using wires as ties. The tightening is accomplished by inserting the ends of the wire into slots at each end of the tightening rod (A, Figure 7-12). The rod is also provided with a slot through which a handle is placed to turn the rod, thus winding the wire until it is brought to the proper tension. The tightener is held tight by inserting a metal clip in the handle slot and against the waler.

ANCHOR BOLTS

An anchor used in building construction is a rod, metal strap, or bolt which secures wood framework to masonry. One end of the metal anchor must be firmly embedded in the solid masonry wall and the other end must be fastened firmly to the sill.

Buildings are anchored to masonry walls to prevent them from shifting from the wall. When a building moves off a wall, there are two distinct movements. First, the building will be lifted vertically and then moved laterally or sideways. This upward movement is caused by the wind getting under the covered roof or sub-floors of a building under construction. This is particularly likely to happen in a building with the platform type of framing in which the sub-floors are laid as the framework is erected to the attic floor line, and the openings in the side walls are not covered. This allows the wind to get under the floors and lift the building.

Buildings are set upon piers extending about two feet above the ground, and, although they are anchored to the piers, it will not take a very high wind to move the building off the piers, once the wind gets under the floors or roof. This danger exists in family dwellings, particularly where the building is two full stories in height. Anchor bolts or straps connecting the building to the piers or foundation walls should be provided but they will be put under a heavy stress if a strong wind gets under the superstructure of the building.

In mass erection of buildings, it is a common practice to frame and roof as many of the buildings

as possible while the weather is good. In bad weather or on rainy days, the carpenters work on the inside of the buildings to finish the partitions, set window frames, and finish as much of the inside work as possible. If this precedure of construction is followed, the skirting should be applied and the window openings of the sheathed side walls should be enclosed on the windward side of the buildings as soon as possible. In no case should the roof be covered before the side walls are enclosed.

The anchor most commonly used in homes and garages is a threaded rod with a nut and washer on one end. The other end is bent to hold in the concrete (Figure 7-13).

Bent Rod Anchor
Figure 7-13

Figure 7-14 shows a straight bolt with nuts and washers and with a piece of pipe placed around the bolt. This prevents the concrete from forming tightly around the bolt, thus allowing it to swivel. This makes fitting sills or equipment bases to the anchor bolts much easier than if the bolts were embedded solidly in concrete.

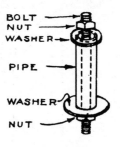

Flexible Anchor
Figure 7-14

A combination anchor and bearing plate used to anchor steel beams to piers is shown in Figure 7-15. The angle clips are bolted or riveted to the steel column, and have holes which coincide with holes in the bearing plate. The rods that are to be embedded in the concrete protrude through the holes in the angle clip and plate and are fastened with a nut as shown.

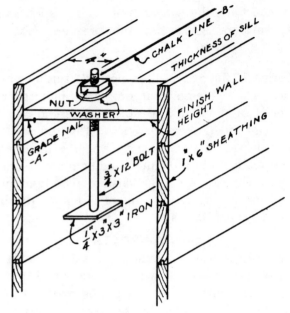

Detail of Suspended Anchor Bolt
Figure 7-17

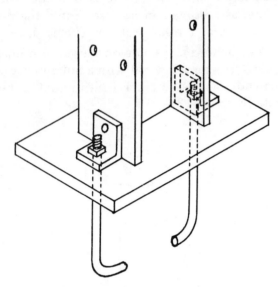

Combination Anchor and Bearing Plate
Figure 7-15

HOW TO INSTALL ANCHOR BOLTS

Anchor bolts which are to hold plates, sills, girders, structural steel, or machinery must be set accurately in the concrete so they will enter the holes in these members. The bolts must also be located accurately in the form and must be kept in place by the use of a wooden templet or pattern in which holes are bored at the proper points to receive the bolts. The templets should be securely fastened to the top of the form.

Assume that a solid concrete foundation wall 12 inches thick is to be poured for a house 26 feet by

50 feet. The forms have been built, aligned, and braced. The anchor bolts are to be set approximately 6 feet apart. Those nearest the corners are to be 1 foot 3 inches from the outside corner of the wall (Figure 7-16). The complete wall will require twenty-eight bolts, nine for each side and five for each end. Nails have been driven in the form about 10 feet apart at the height of the finished concrete wall. These are called grade nails or points. See A, Figure 7-17.

1. Snap a chalk line from one grade nail to the next, making a continuous line around the form (B, Figure 7-17).

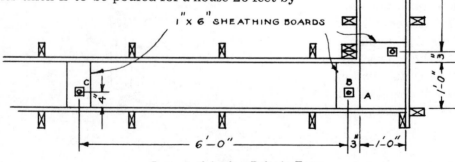

Layout of Anchor Bolts in Form
Figure 7-16

60

2. Cut 28 pieces of sheathing 1 inch x 6 inches x 12 inches. These are to be placed in the form to suspend the anchor bolts until the concrete has set.

3. Nail the first piece in place with its center line 1 foot 3 inches from the outside corner of the wall. The under side of the piece should be at the same height as the chalk line (B, Figure 7-17).

4. Locate the first anchor bolt hole by measuring in 4 inches from the outside face of the wall along the center line of the piece of sheathing. Drive a nail temporarily at this point (Figure 7-16 at B).

5. Hook the ring of a steel tape over the nail. Run the tape out and mark the form every 6 feet; that is, mark at 6 feet, 12 feet and 18 feet. These points mark the center lines of the 12 inch boards.

6. Locate and fasten the boards at these points in the manner described in Step 3.

7. Locate the hole in each board as described in Step 4.

8. Bore holes at these points, 1/16 inch larger than the anchor bolt that is to be used.

9. Slip the bolts in place and run the nut and washer down on the bolt until it projects above the concrete line the thickness of the sill plus about 1/2 inch.

NOTE: When the concrete is being poured, be sure that it does not push the bolts out of place.

ANCHOR BOLTS FOR ENGINE BEDS AND STEEL COLUMNS

NOTE: Figure 7-18 shows a method of using a wooden templet to suspend a group of anchor bolts where accuracy is important. An example of this would be in setting anchors for an engine bed or for steel columns. If the templet is made as a complete unit and installed on the form as such, it should be squared and nailed together first, so it will remain square until secured in position.

To locate flexible anchors, proceed as in locating anchors for columns and engine beds.

NOTE: The washer to be embedded in the concrete should have a diameter of at least 2 inches greater than that of the pipe used around the bolt.

Place a washer of approximately the same size as the outside diameter of the pipe, on the upper end of the bolt and draw it down tightly against the pipe with the nut. This prevents the wet concrete from getting into the inside of the pipe (A, Figure 7-19).

Suspend the bolts in the templet so that the top edge of the pipe will be level with the top of the finished concrete work (B, Figure 7-19).

Figure 7-20 shows a combination anchor bolt and bearing plate. These plates come in many different styles, some being welded together and some bolted. However, the procedure in locating them in the concrete is similar to that used for the other type described.

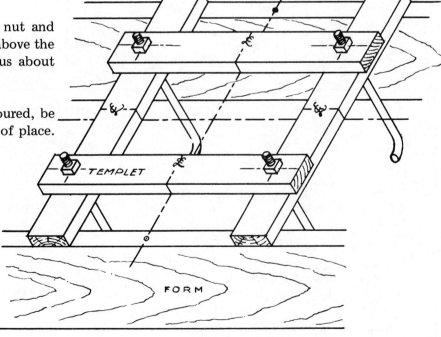

Anchor Bolts Held in Position by Templet
Figure 7-18

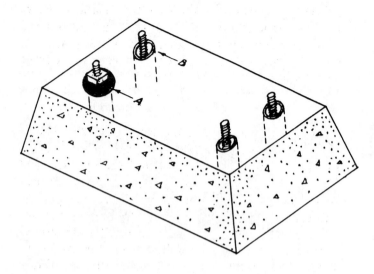

Flexible Anchor Bolts in Place
Figure 7-19

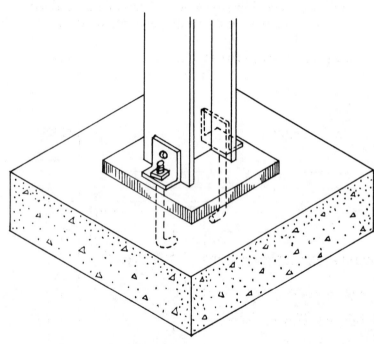

Anchor and Bearing Plate in Position
Figure 7-20

CONSTRUCTION JOINTS

Purpose. Construction joints are used between the units of a structure and located so they will not cause weakness. It would be preferable theoretically that each beam, girder, column, wall, or floor slab be placed in one operation to produce a homogeneous member without seams or joints, but for practical reasons, this procedure is usually impossible. The planes separating the work done on different days, called construction joints, are placed where they will cause the minimum amount of weakness to the structure, and where the shearing stresses and bending moments are small or where the joints will be supported by other members.

Construction Joints Between Wall and Footing. At a construction joint between a wall and its footing, a keyway is usually necessary to transfer the shearing stresses. A keyway must always be used if no reinforcing steel or dowels tie the wall and footing together. This keyway can be formed by pressing a slightly beveled 2 by 4 into the concrete before it has set and removing the 2 by 4 after the concrete has hardened. The 2 by 4 should be well oiled before it is used. Such a keyway is shown in figure 7-1. If the wall and footing can be placed at one time, a construction joint is not necessary.

Vertical Construction Joints. If it is desirable to deposit the concrete for the full wall height, the forms should be divided into sections by vertical bulkheads as shown in Figure 7-21. These bulkheads should be spaced so that the complete section can be filled in one continuous operation. Experience has shown that the V-joint in Figure 7-22 is less likely to break off than the joint shown in Figure 7-21. If reinforcing steel or dowels cross the joint no projection is needed.

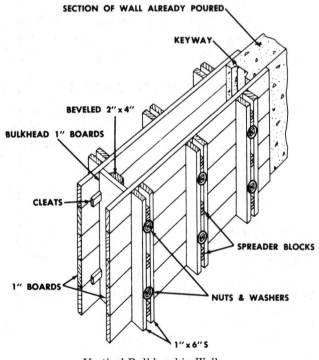

Vertical Bulkhead in Wall
Figure 7-21

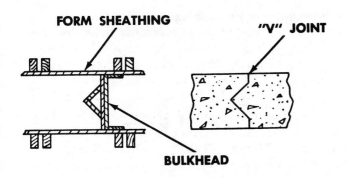

FORM SHEATHING

"V" JOINT

BULKHEAD

Vertical V-Construction Joint
Figure 7-22

Beam, Column, and Floor Slab Joints. The proper place to end a pour in construction involving beams, columns, and floor slabs is shown in Figure 7-23. The concrete in each column should be placed to the underside of the beam or floor slab above. A construction joint in a beam or floor slab should occur at the center of the span so as to avoid points of maximum shear. All construction joints in beams and slabs should be vertical. Reinforcing steel or dowels should extend across the joint. A beam or slab should never be placed in two lifts vertically, for this produces a weak joint between the two layers.

EXPANSION AND CONTRACTION JOINTS

Principle. Shrinkage of concrete during hydration is comparable to a drop in temperature of 30° to 80° F. depending on the richness of the mix. Contraction joints are necessary to permit the concrete to shrink during the curing process without damage to the structure. The structural engineer will consider the best place for the joints from a standpoint of serving the purpose. They are usually placed where there is a change in thickness, at offsets and where the concrete will tend to crack if shrinkage and deformations due to temperature are restrained. Joints should be about 30 feet center-to-center in exposed structures.

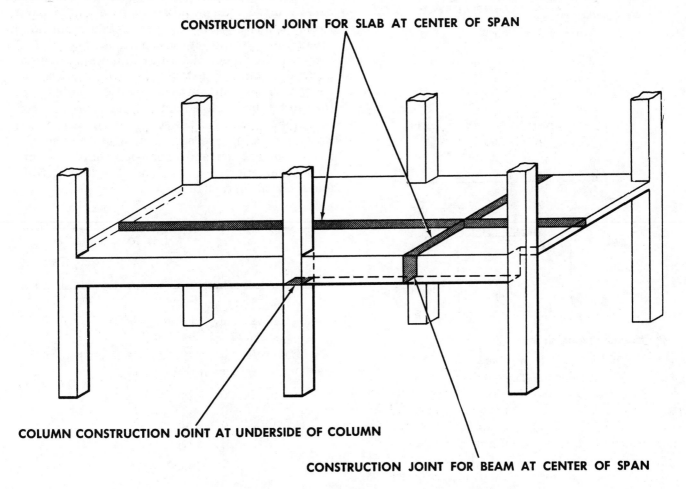

CONSTRUCTION JOINT FOR SLAB AT CENTER OF SPAN

COLUMN CONSTRUCTION JOINT AT UNDERSIDE OF COLUMN

CONSTRUCTION JOINT FOR BEAM AT CENTER OF SPAN

Location of Construction Joints in Beams, Columns, and Floor Slabs
Figure 7-23

Expansion Joints. Expansion joints, in the form of vertical joints through the concrete, are used to isolate adjacent units of a structure, to prevent cracking due to shrinkage or temperature changes. Generally, an expansion joint is used with a premolded mastic or cork filler, if an elongation of adjacent parts and a closing of the joint is anticipated. Expansion joints for different types of structures are illustrated in Figures 7-24 through 7-26. Expansion joints should be installed every 200 feet.

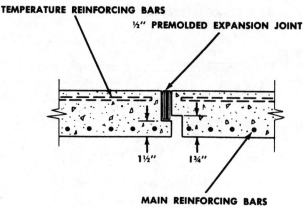

Expansion Joint for Floor Slab
Figure 7-26

Contraction Joints. The purpose of contraction joints is to control cracking due to temperature changes incident to shrinkage of the concrete. If the concrete cracks it will usually occur at these joints. Usually, the contraction caused by shrinkage will offset a large part of the expansion due to a rise in temperature. Contraction joints are usually made with no filler or with a thin paint coat of asphalt, paraffin, or some other material to break the bond. Joints as shown in Figure 7-27 should be installed at 30-foot intervals or closer depending on the extent of local temperature change. These dummy contraction joints are formed by cutting to a depth of one-fourth to one-third the thickness of the section. Contraction and expansion joints are not used in beams and columns. Contraction and expansion joints in reinforced concrete floor slabs should be placed at 100-foot intervals in each direction.

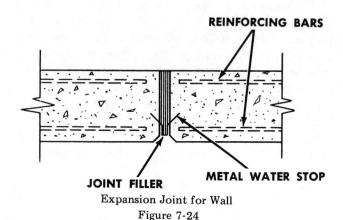

Expansion Joint for Wall
Figure 7-24

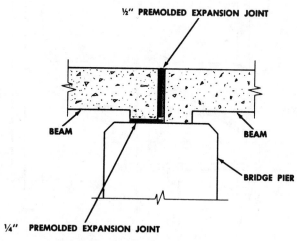

¼" PREMOLDED EXPANSION JOINT

Expansion Joint for Bridge
Figure 7-25

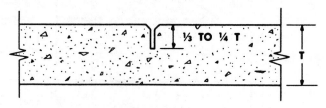

Dummy Construction Joint
Figure 7-27

Concrete Pavement Joints. Concrete pavement joints as shown in Figure 7-28.

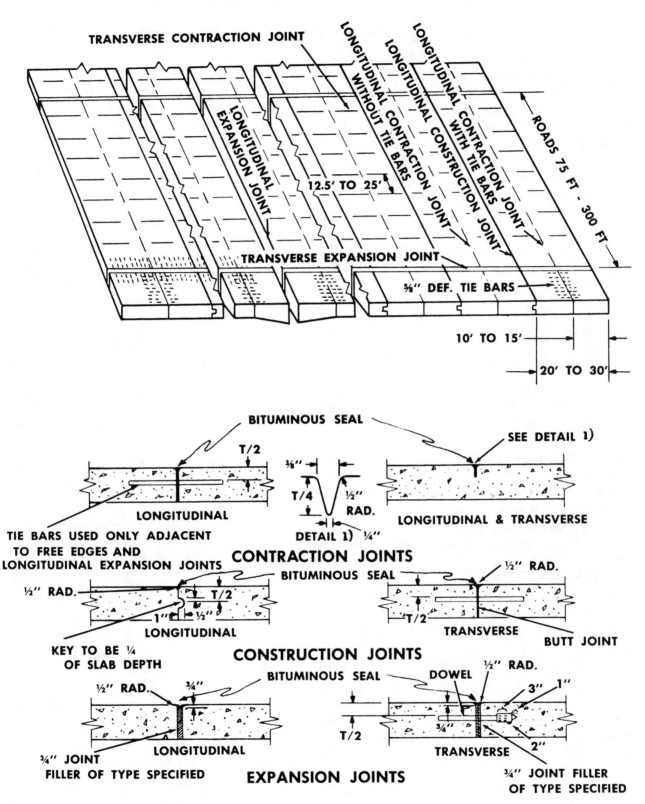

TRANSVERSE CONTRACTION JOINT

LONGITUDINAL CONTRACTION JOINT

LONGITUDINAL CONTRACTION JOINT WITH TIE BARS

LONGITUDINAL CONSTRUCTION JOINT

LONGITUDINAL CONTRACTION JOINT WITHOUT TIE BARS

LONGITUDINAL EXPANSION JOINT

ROADS 75 FT - 300 FT

12.5' TO 25'

TRANSVERSE EXPANSION JOINT

⅝" DEF. TIE BARS

10' TO 15'

20' TO 30'

BITUMINOUS SEAL

T/2

SEE DETAIL 1)

TIE BARS USED ONLY ADJACENT
TO FREE EDGES AND
LONGITUDINAL EXPANSION JOINTS

LONGITUDINAL

⅜"

T/4   ½"
       RAD.

DETAIL 1) ¼"

LONGITUDINAL & TRANSVERSE

CONTRACTION JOINTS

½" RAD.

BITUMINOUS SEAL

T/2

1"   ½"

LONGITUDINAL

KEY TO BE ¼
OF SLAB DEPTH

BITUMINOUS SEAL   ½" RAD.

T/2

TRANSVERSE

BUTT JOINT

CONSTRUCTION JOINTS

½" RAD.   ¾"

BITUMINOUS SEAL

LONGITUDINAL

¾" JOINT
FILLER OF TYPE SPECIFIED

DOWEL   ½" RAD.

BITUMINOUS SEAL

3"   1"

T/2

¾"

TRANSVERSE   2"

¾" JOINT FILLER
OF TYPE SPECIFIED

EXPANSION JOINTS

Design and Spacing of Joints for Concrete Slab
Figure 7-28

CHAPTER 8

CONSTRUCTION OF PIER AND FOOTING FORMS

A pier is a post or column extending into and above the ground and supporting a structural member of a building. It may be made of wood, metal, or concrete, the size and shape being governed by the weight of the structure it is to support.

Footings for piers and walls are made in many different ways, the size, design and construction depending on the load to be carried and the bearing strength of the soil. The fundamental purpose of a footing is to distribute the load imposed on the pier over a larger area of ground than would be covered by the pier or column itself.

A footing should be made strong enough to support the dead weight of the wall or pier and also the live and dead weight imposed on the wall. In general, a footing for a foundation wall or pier for light frame construction should extend no more than 4 inches beyond each side of the wall or pier, unless the footing is reinforced or tapered.

The thickness of the footing should be 10 inches to 12 inches. If keyways are used, they should not impair the strength of the footing. That portion of footings and piers which is above the frost line should be tapered slightly toward the top so that the heaving of the soil will not tend to heave the pier.

Once frost gets under footings it will raise them. This is particularly true of small light buildings such as garages and houses. It is doubtful whether frost could raise a masonry building which is two or three stories high. Nevertheless, the footings of such buildings are placed below the frost line.

If the building site slopes, and if the excavation is dug parallel to this slope and the footings are poured accordingly, there will be a tendency for the building to "creep" or slide to the lowest point. This may be overcome by stepping the excavation to meet the contour of the ground. The thickness of one footing that laps over another should not be less than 8 inches. See Figure 8-1.

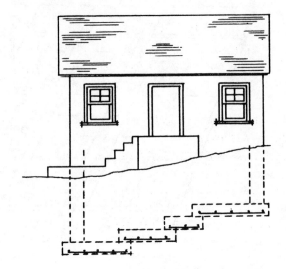

Stepped Footing
Figure 8-1

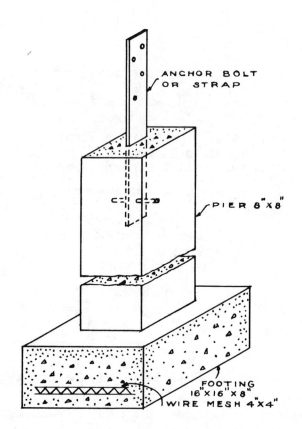

Pier and Footing Form
Figure 8-2

Footings may be classified into two general groups; pier or independent footings, and wall or continuous footings.

PIER FOOTINGS

Pier or column footings may have stepped or tapered sides. Figure 8-2 shows the type of pier used for ordinary garage and light frame construction.

Figure 8-3 shows the stepped footing generally used where the load on the pier per square foot is greater than the bearing strength of the soil per square foot. Figure 8-4 shows the tapered footing which is also used to distribute the load over a large area. The sides are tapered to conserve material and to prevent frost from heaving the footing.

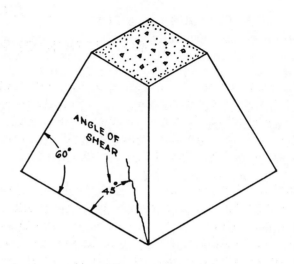

Tapered Footing
Figure 8-4

Combination wall and column footings are used when a wall and a column are combined to form a pilaster (Figure 8-5). This type of footing is used where piers are set at intermediate points in a continuous wall such as at A, Figure 8-5, or where there is a concentrated load on the wall.

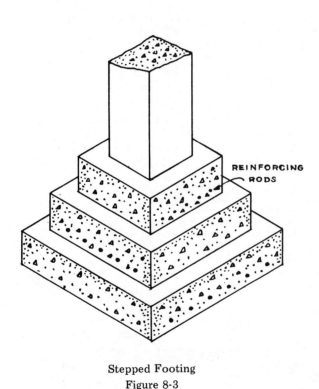

Stepped Footing
Figure 8-3

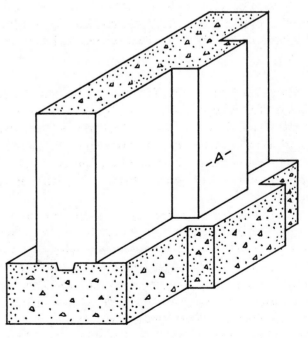

Pilaster Wall Footing
Figure 8-5

The angle of slope used for tapered pier footings should be 60 degrees from the horizontal. This is based on the fact that the angle of shear caused by a load on the footing is 45 degrees. This angle is well within the limits of the 60 degrees sloped sides of the pier footing.

CONTINUOUS WALL FOOTINGS

Footings are generally shown on foundation plans by dotted lines on both sides of the wall (Figure 8-6). These lines represent the outside and inside edges of the footings. The section through the wall at A-A shows the size of the footing. In this case the wall is one foot thick with a 4 inch projection on each side of the wall, making the footing 1 foot 8 inches wide. The thickness or depth of the footing is 1 foot.

It is not always necessary to build footing forms, particularly where the earth is solid or firm. In this case, a trench may be excavated to the width and depth of the footing and concrete poured into this trench (Figure 8-7). This saves considerable time and material, but if the soil is such that it will not stand up firmly and is porous it will be necessary to build forms.

FOOTING FORMS

Concrete footing forms (Figure 8-8) are usually constructed of 2 inch planks set on edge. The width of the plank is the same as the depth of the footing. These planks are held in place with stakes and are held the correct distance apart with cross-spreaders or ties. At each stake, a short brace holds the forms in line. Due to irregularities in the excavation, it may be necessary to trim off dirt in some places to get the footing form level. In other places, the form may have to be raised. Dirt is then thrown around the outside of the form to prevent the wet concrete from running out. The concrete is then poured into the forms and screeded off level with the top of the planks.

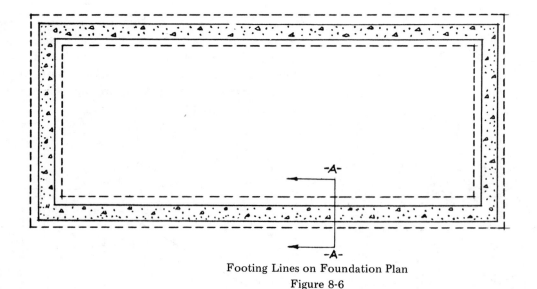

Footing Lines on Foundation Plan
Figure 8-6

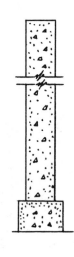

Section A-A

Trench to act as Form for Concrete Footing
Figure 8-7

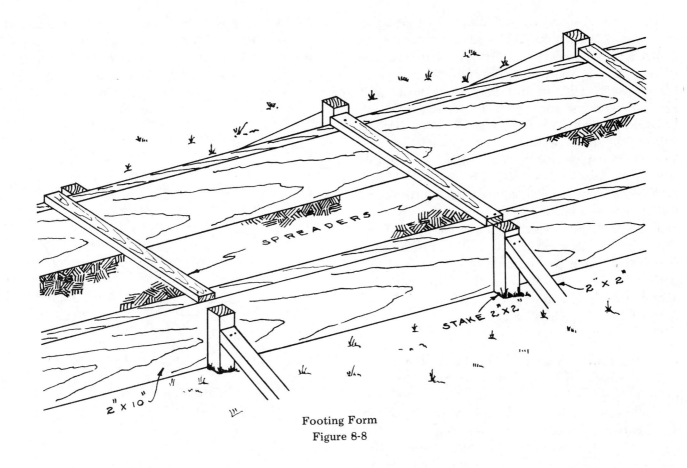

Footing Form
Figure 8-8

Another type of form is used where it is impossible to drive stakes to hold the forms in place. This will occur where the foundation is hardpan or rock. In this case, the planks used are wider than the depth or thickness of the footing. The top of the forms are not set to any definite level, but are set to meet the contour of the ground.

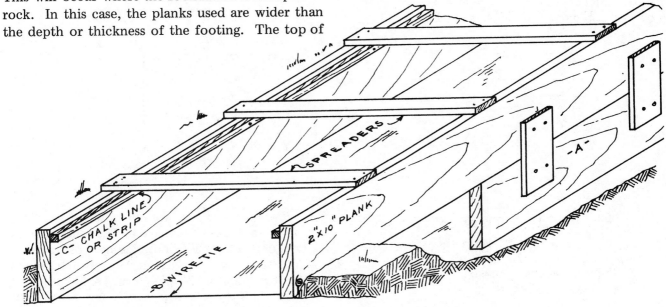

Footing Form for hard irregular ground
Figure 8-9

The form is made by fastening 2 x 4 cleats to the sides of the planks and building the form sides higher where necessary to follow the contour of the ground. See A, Figure 8-9. Short braces are placed against each cleat, or dirt may be piled at each cleat to hold the form in place. The planks are held in place with a tie at the top and one directly below. In some cases, it is not possible to have wood spreaders at the bottom of the form. In this case, wires may be used instead of wood (B, Figure 8-9). Because of the difficulty of making the forms level, they are made higher than necessary, and nails or strips are placed on the inner sides of the form at the elevation of the top of the footing. Concrete is then poured and screeded off at the strip or nails. This method is the quickest and easiest for the carpenter but causes some difficulty for the concrete gang (C, Figure 8-9).

BONDING A WALL TO A FOOTING

Some provision is necessary for tying the wall above to the footing below, particularly where one side acts as a retaining wall (Figure 8-10). If no such provision is made, there will be a tendency for the wall to be pushed off the footing due to the pressure of the earth on one side with nothing on the other side to counteract it. A keyway in the center of the footing will prevent this. The keyway may be made either by pressing a slightly beveled 2 x 4 into the concrete while it is still wet and then removing it when the concrete has set (Figure 8-11),

or by pressing hard bricks into the footing and allowing about one half to project above the top of the footing (Figure 8-12). Iron dowels or reinforcing rods may also be used to fasten a wall to a footing. In a case where the footing and the wall can be poured together, the keyway or rods are not necessary.

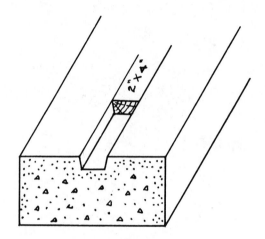

Methods of making Keyway
Figure 8-11

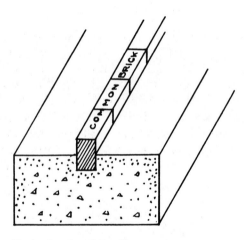

Methods of making Keyway
Figure 8-12

Pier Forms. Forms for concrete piers are made of several types of materials. The most common type of form is made of tongue and groove sheathing, nailed together with cleats to form a rigid panel for a side of the pier form (A, Figure 8-13). Four sides are made in this manner, the width and length being dependent on the size of the concrete pier to be constructed. These four sides are then nailed together in the form of a square

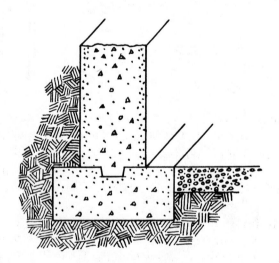

Footing Bonded to Wall
Figure 8-10

and are held together by yokes or column clamps shown at B and C, Figure 8-13. These yokes prevent the wooden form from bulging when the concrete is poured, and hold the form in shape until the wet concrete has set. This type of form is used for columns or piers less than 5 feet high and not over 2 feet square. Where footings are required underneath the pier, box like frames of the size of the footing are made.

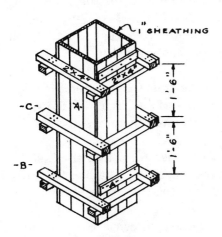

Form for short Column
Figure 8-13

A form that is sometimes used is shown in Figure 8-14. It measures 10 inches to 12 inches each way on the top and is somewhat larger at the bottom so it can be lifted off the concrete. The top level of the pier can be regulated by putting blocks under the handles of the form. By digging the hole for the base of the pier about 6 to 8 inches larger each way than the form, provision may be made for forming the footing of the pier.

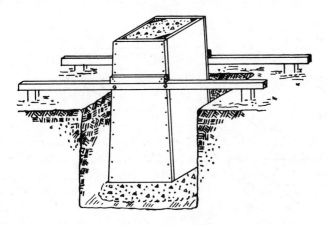

Form for tall Column
Figure 8-14

Column and post forms that are filled with concrete in an upright position are subjected to greater pressure near the bottom of the form, therefore, they must be built more strongly at the bottom than at the top. The easiest way to do this is to put the yokes or clamps more closely together near the bottom of the form. In column forms, the boards should be placed vertically and should be tongued and grooved so the pressure of the concrete will not force water and cement between the boards (Figure 8-15).

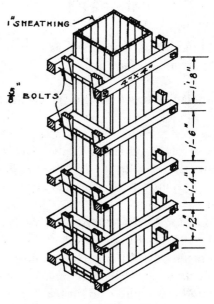

Removable Pier Form
Figure 8-15

Large, heavy columns must have stronger forms because of the great weight of the concrete at the bottom (Figure 8-15). These forms have yokes of 4 x 4 material held together by bolts, and side pieces held in position by wedges.

The forms may be taken off a concrete column that is to carry no weight, within 24 to 48 hours after the concrete has been poured into the form. Forms for load bearing piers should be left on 48 to 96 hours.

71

BUILDING PIER AND FOOTING FORMS

In some cases, the footing forms and the pier or wall forms are each set and poured separately. In this type of construction, the two masses of concrete must be bonded together. Another method of building this footing and pier is to construct the form in such a manner that the footing and pier can be poured together. Either of two methods might be used to accomplish this. Separate forms may be built and set together, or only the pier form may be used and the footing formed by the enlargement of the bottom of the excavation.

STEPS IN MAKING SMALL PIER FORMS (FIGURE 8-14)

1. Cut the four sides to the desired width from solid wide boards and taper them as shown. Apply grease to the inside of each panel.

2. Nail the sides together and fasten the handles in place.

NOTE: If plywood is used it will be necessary to reinforce the corners with cleats and nails as nails alone will not hold them securely enough.

HOW TO MAKE COMBINATION FOOTINGS AND PIER FORMS

1. Rip a piece of sheathing of the required length to 2¾ inches wide. Match this piece with 5¼ inch wide sheathing board and fasten them together with cleats as shown in Figure 8-16A. Allow the ends of the cleats to project 3/4 inch. Make another panel of the same size. One of these panels is shown at A, Figure 8-17. The other one is on the opposite side of the form.

2. Rip two pieces of sheathing 4¼ inches wide and match each piece with a 5¼ inch wide sheathing board. Build two panels as described in Step 1 and as shown in Figure 8-16B. These are to be used as shown at B, Figure 8-17.

3. To assemble the four panels, butt them together and nail as shown in Figure 8-17.

4. Make a framework for the footing by cutting two pieces of 1-5/8 inch x 7-1/2 inch as at C, Figure 8-17, and two pieces as at D. Nail these pieces together as shown.

5. Square the frame and fasten pieces of sheathing on two sides of the frame as shown at E. Cut two pieces and fasten them as shown at F.

6. Insert the pier form into the hole left in the top of the footing form. The pier form will be supported on the footing form by the cleats (G).

7. Make yokes as shown at H and I and place one such yoke every two feet on the pier form.

8. If anchor bolts are to be used, fasten a templet (J) to the top of the form.

HOW TO MAKE INDIVIDUAL PIER FOOTING FORMS

1. Figure 8-18 shows a form made of 1 inch stock. To make the sides A and B, cut six pieces to the required width and length to correspond to the thickness and length of the concrete footing.

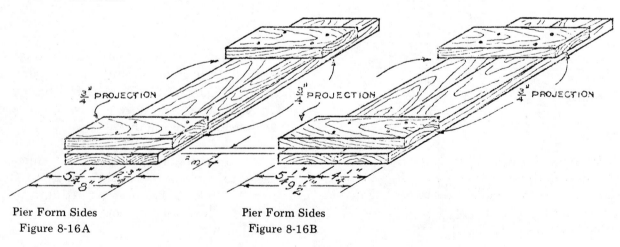

Pier Form Sides
Figure 8-16A

Pier Form Sides
Figure 8-16B

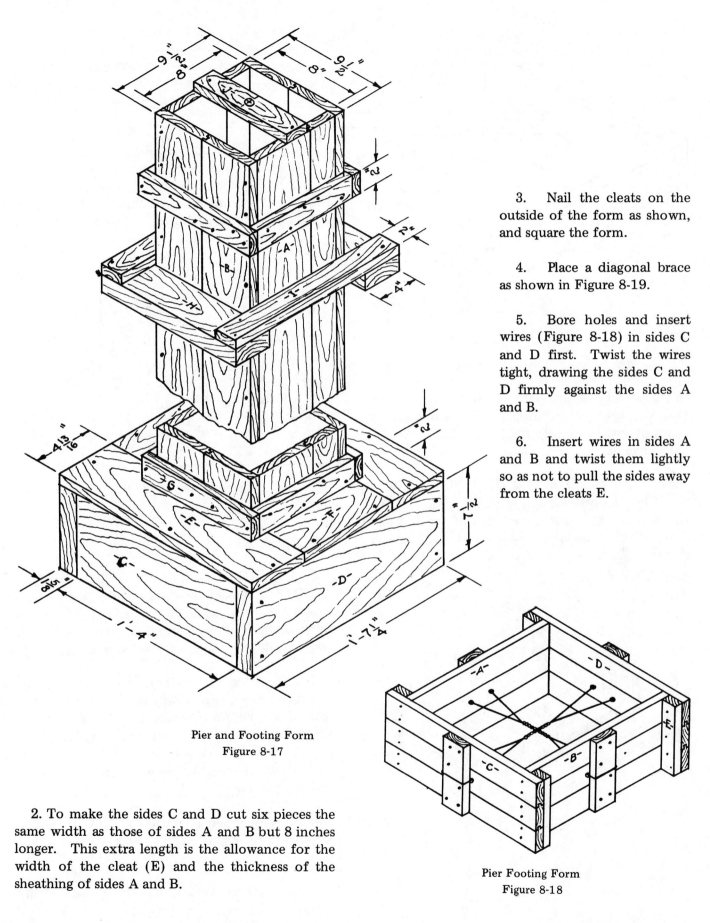

3. Nail the cleats on the outside of the form as shown, and square the form.

4. Place a diagonal brace as shown in Figure 8-19.

5. Bore holes and insert wires (Figure 8-18) in sides C and D first. Twist the wires tight, drawing the sides C and D firmly against the sides A and B.

6. Insert wires in sides A and B and twist them lightly so as not to pull the sides away from the cleats E.

Pier and Footing Form
Figure 8-17

Pier Footing Form
Figure 8-18

2. To make the sides C and D cut six pieces the same width as those of sides A and B but 8 inches longer. This extra length is the allowance for the width of the cleat (E) and the thickness of the sheathing of sides A and B.

73

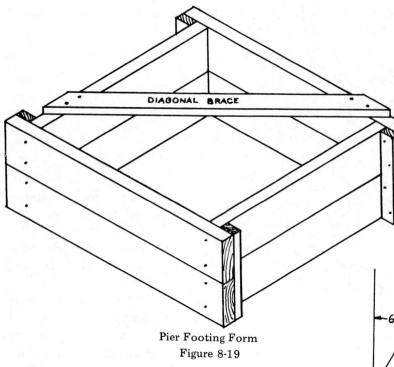

Pier Footing Form
Figure 8-19

Figure 8-19 shows a form made with 2 inch stock. This type of form requires no wires or cleats because the heavy stock is rigid and provides better nailing.

HOW TO MAKE STEPPED FOOTING FORMS

1. The stepped form is a series of individual forms made to the required size of the footing. The individual parts are made similar to the form shown in Figure 8-19. The complete form is shown in Figure 8-20.

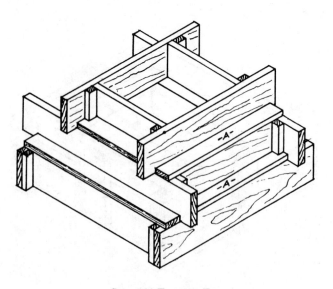

Stepped Footing Forms
Figure 8-20

2. Place sheathing boards over the openings at the top of each step of the form as at A, Figure 8-20.

3. Figure 8-21 shows a section of the stepped form with one top sheathing board in place. If these boards are not provided, the resulting shape of the concrete pier will be as shown.

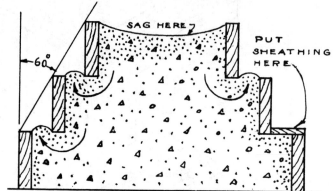

Section of Stepped Form
Figure 8-21

HOW TO MAKE TAPERED FOOTINGS

NOTE: To make the sides A and B, Figure 8-22, proceed in the manner described for making the sides of the form in Figure 8-18 except that allowance must be made for the taper of the sides.

1. Match the required number of sheathing boards together, and lay off the width of the bottom of the pier on board A, Figure 8-23. Lay off the width of the top of the pier on board B. Be sure to work from a center line.

2. Mark off straight lines connecting the bottom and top points.

3. Cut along these lines and nail the cleat in place as shown in Figure 8-23. Make two of these panels.

4. To make the sides C and D, Figure 8-22, fol-

74

low the same procedure, making allowances so these sides will fit against sides A and B. Nail the three cleats as shown in Figure 8-24.

5. Assemble and brace the form as shown in Figure 8-18 and 8-19.

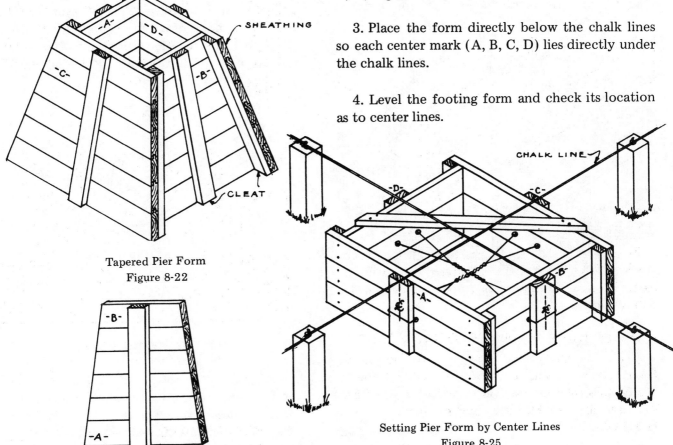

Tapered Pier Form
Figure 8-22

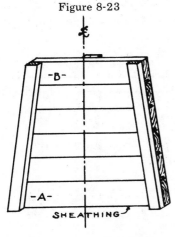

Panel of Tapered Pier Form
Figure 8-23

Panel of Tapered Pier Form
Figure 8-24

HOW TO SET AND BRACE FOOTING FORMS

1. Locate the two center lines of the proposed pier footing by stretching chalk lines on stakes as shown in Figure 8-25.

2. Mark the center of the top of each side (A, B, C, D, Figure 8-25).

3. Place the form directly below the chalk lines so each center mark (A, B, C, D) lies directly under the chalk lines.

4. Level the footing form and check its location as to center lines.

Setting Pier Form by Center Lines
Figure 8-25

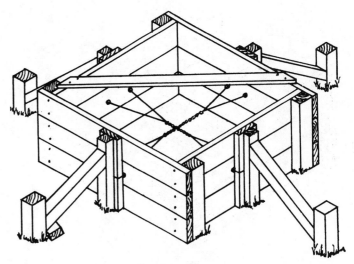

Bracing a Form
Figure 8-26

5. Brace the form as shown in Figure 8-26 or pack earth around it as shown in Figure 8-27.

6. Locate and brace pier forms, such as shown in Figure 8-17, in a similar manner, but place additional braces against the column yokes (H) and extend them to stakes driven into the ground.

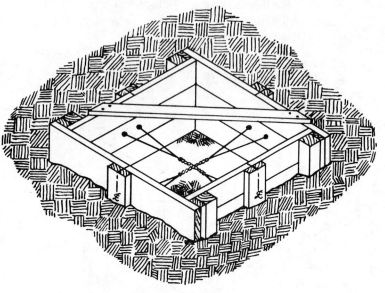

Form Set in Ground
Figure 8-27

HOW TO BRACE STEPPED OR TAPERED FOOTING FORMS

1. To brace this type of form, use side bracing similar to that used for the other footing forms.

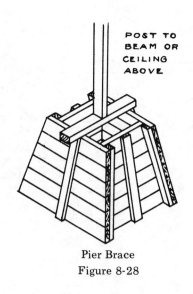

Pier Brace
Figure 8-28

2. To brace the form against heaving or floating by the method shown in Figure 8-28, cut the upright and crosspiece to length as shown, and toe nail the upright to a floor beam and to the crosspiece. Then nail the crosspiece to the top of the form.

3. To brace the form against heaving or floating by the method shown in Figure 8-29, nail planks along two sides of the form. Nail two cross planks on top of these, forming a platform on which heavy weights may be placed.

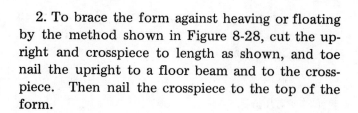

Pier Anchor
Figure 8-29

4. Figure 8-30 shows another method of anchoring a tapered form. Partly fill the form with concrete. Place a wire anchor in the concrete and after the concrete has hardened twist the wire around the crosspiece at the top of the form.

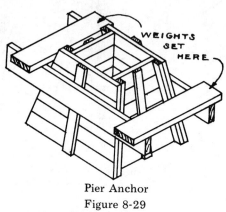

Tapered Form Anchor
Figure 8-30

HOW TO MAKE AND SET FORMS FOR WALL FOOTINGS

1. To set the footing forms in the proper relation to the building lines, place chalk lines on the batter boards as they were during the layout of the excavation.

2. At the corners of the building line, shown by the crossing of the chalk lines, plumb down into the excavation to locate the outside corner of the wall. Measure from this point to locate the stakes (B and C, Figure 8-31), marking the outside and the inside corners of the footing, allowing for the thickness of the form material and the projection of the footing beyond the face of the wall.

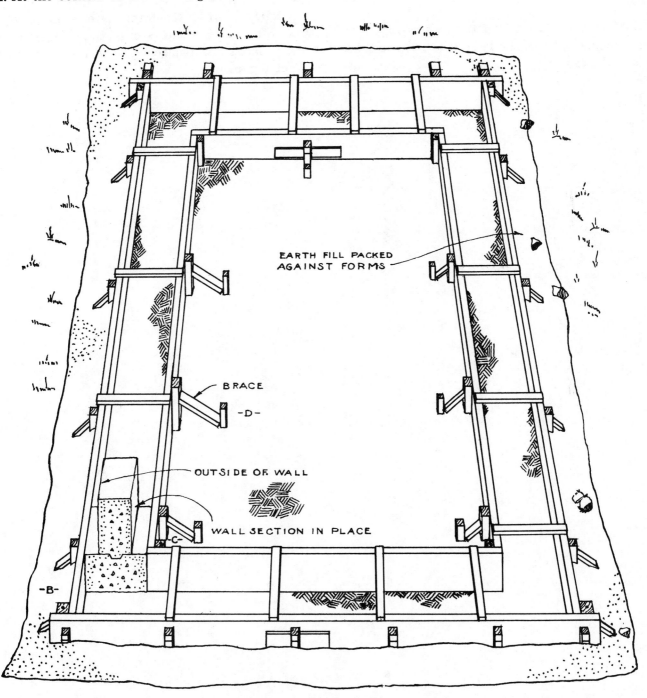

EARTH FILL PACKED AGAINST FORMS

BRACE
-D-

OUTSIDE OF WALL

WALL SECTION IN PLACE

-C-

-B-

Footing Forms Set in Place
Figure 8-31

3. Drive the top of each stake (B and C) down to the level of the top of the proposed footing.

4. Repeat this operation at each corner of the excavation. Drive a nail in the top of each stake.

5. Stretch chalk lines from these nails, to show the alignment of the inside and outside footing forms (Figure 8-31).

6. Make the side walls of the footing forms, using 2 inch planks of the same width as the thickness of the footing.

7. Nail 2 inch strips across the top of the form about every 3 feet. Drive stakes into the ground against the outside of the form to keep the bottom of the form from spreading.

8. Drive stakes (D, Figure 8-31) firmly into the ground about 12 inches from the sides of the form and insert the braces.

NOTE: After the forms have been braced and earth has been packed around the outside of them, they should be rechecked as to levelness and correct alignment.

CHAPTER 9

CONSTRUCTION OF FOUNDATION WALL FORMS

There are two general systems used in making forms for concrete walls. The less expensive of the two systems requires only an inside form. The complete or double form consists of form panels built for both sides of the wall.

Inside Wall Forms. When the foundation walls come to the limits of the lot line, it is necessary to build only an inside form. The bank of the excavation serves as the outer form. However, this type of form should be avoided whenever possible, as it is very difficult to cut the banks straight and plumb. If the banks are not straight and plumb, it is necessary to use more concrete, which makes the wall more expensive. Also, it is difficult to waterproof this type of wall from the outside.

Without an outside form, no wall ties can be used. This type of construction requires heavier bracing on the inside wall to compensate for this lack of wall ties.

The concrete should be poured in slowly and not from a great height as the higher the concrete fall, the greater the pressure on the forms. All the inside surfaces of the form should be painted with a stainless grease, crude oil, soft soap, or white wash to prevent the concrete from sticking to the form.

Complete Inside And Outside Forms. When the foundation walls are well within the lot or property lines, forms for both the inside and outside of the wall should be used. With this type of form, a reasonably straight smooth wall is possible. When double forms are used, the excavation must extend beyond the actual building line. If the cellar is dug at least 12 inches outside each wall line, it will provide enough space for the outside form and also space to waterproof the wall with tar or other materials.

Forms should be built in sections that are not too heavy for four men to lift, and they should be erected and secured in such a way that they may be removed from the hardened concrete wall without an excessive amount of hammering or pulling which might crack or deface the concrete.

After the form sections have been made, they should be braced in position by walers and spreaders, and held to alignment by diagonal bracing. See Figure 9-1. Wire is often used to tie the forms together at the spreaders. The spreaders should be removed as the concrete is poured into the forms. Otherwise the wooden spacers will form a blemish in the finished wall. Care and judgment should be used when twisting the wires in a form so as not to break them. Too much twisting will stretch the wire to a point close to its tensile strength, and the wire is likely to break under the pressure of the concrete load.

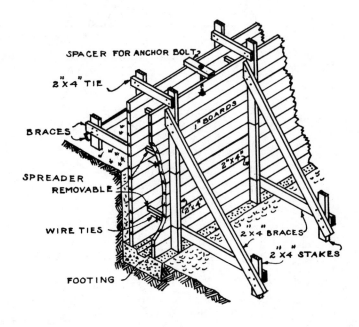

Double Wall Form
Figure 9-1

HOW TO MAKE PANELS FOR SINGLE WALL FORMS:

NOTE: If the wall is very long, make up the forms in panels, not over 10 feet in length for easy handling.

1. Cut 2 x 4 studs long enough to reach the top

of the proposed form.

2. Place two 12 foot planks on top of a pair of saw horses. The planks may be 2 x 6 or larger.

3. Spread the planks on the horses so that the studs will lie across the planks.

4. Nail the two outside planks in place on the saw horses to keep them from moving.

5. Select a straight sheathing board and cut it to the length of the panel.

6. With a steel square, measure 24 inches from the left end of the board and make a mark. Square this mark across the board and make an X mark at the right of it (Figure 9-2).

7. Using the body of the framing square, continue to lay out the location of the studs at 24 inch intervals across the board. Let the last stud come as it will. The marked board should look like A, Figure 9-2.

8. Make a duplicate of this board as shown at B.

9. Place studs to correspond to the marks on the boards A and B.

10. Lay the board (A) on top of the studs with the marks up.

11. Fasten the board (A) to the stud on the left with one nail, placing the end of the board flush with the face of the stud (Figure 9-2).

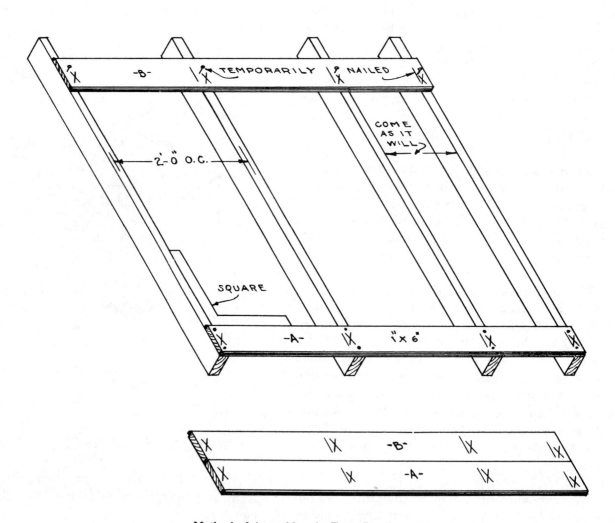

Method of Assembly of a Form Panel
Figure 9-2

NOTE: If a shoe is to be used at the bottom of the form, nail the board (A) to the bottom end of the studs instead of to the face of the studs as shown. Nailing a shoe on the end of the studs makes the panel more rigid and also simplifies the aligning of the panels on the footing.

12. Nail the other end of the board (A) in the same manner.

13. Nail the board to the remaining studs, after having placed them on the side of the line marked X.

14. Place the tongue of the square on the upper edge of the board with the body up the side of a stud, and move the stud until it is square with the board. Nail the board and stud solidly together. Continue in this manner until all studs have been nailed square with the first sheathing board.

15. Tack board (B) temporarily to the top of the studs to hold them in position.

HOW TO NAIL THE SHEATHING

1. Nail the sheathing to the studs, keeping all boards tight and using 6d common wire nails.

2. Nail each sheathing board on every other stud as shown in Figure 9-3.

3. Nail only the groove edge of the sheathing boards on the inside studs. Double nail the top and bottom boards and also the ends of every board.

4. If a joint occurs in the sheathing, place a nail directly over and below the joint on the adjacent boards.

HOW TO SET THE FORMS ON THE FOOTINGS

1. Using a plumb bob, mark points on the footing directly below the building line or outside face of the wall (Figure 5-3).

2. From these points, locate the inside of the wall on the footing.

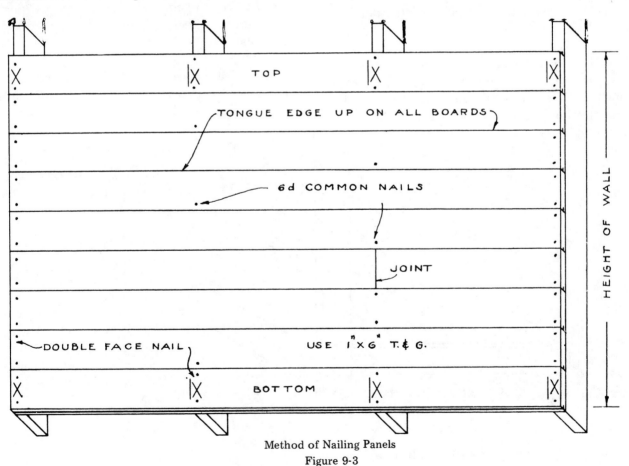

Method of Nailing Panels
Figure 9-3

81

3. At these points, snap a chalk line on the footing. These lines should run the full length of the footing.

4. On this line, drive 8d nails part way in the footing. The nails should be about 6 feet apart.

5. Set the form panels in place against these nails.

6. Brace the bottom of the form to the footing by driving stakes into the ground along the outside edge of the footing as shown in Figure 9-1.

7. Brace the sides of the form panels temporarily.

8. Proceed in similar fashion, erecting each form.

9. Place the walers along the sides of the form panels and place the diagonal side braces about 4 feet apart.

10. Align the panels and plumb them opposite each brace.

NOTE: It is sometimes good procedure to embed a plank in the ground and pack earth firmly around it. The diagonal braces are then nailed against the face of the plank as shown in Figure 9-4.

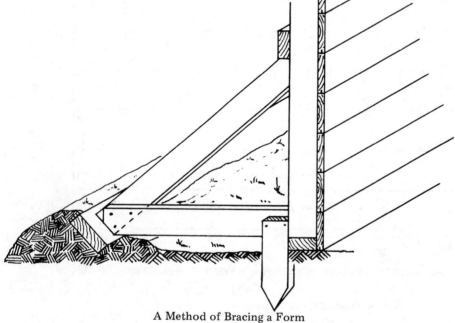

A Method of Bracing a Form
Figure 9-4

HOW TO MAKE AND SET COMPLETE SINGLE WALL FORMS

NOTE: Assume that panel forms are to be built for a building 20 feet wide and 100 feet long, the wall being 12 inches thick and 4 feet high. The inside dimensions of the wall would be 18 feet by 98 feet.

1. Make twenty panels, as previously described, 10 feet long and 4 feet high. Nine of these will be used for each side of the excavation and two for the ends.

2. Make four panels 8 feet long and 4 feet high. One of these will be used for each end and one for each side.

3. Set the panels on the footing and brace and align them as previously described.

HOW TO MAKE PANELS FOR DOUBLE WALL FORMS

1. To build a panel 10 feet long and 4 feet high, (or the height of the wall) cut two pieces 2 inches x 4 inches x 10 feet. These are for the top and bottom plates.

2. Cut six pieces 2 inches x 4 inches x 3 feet 8¾ inches. These are for the uprights or studs (Figure 9-5).

3. Nail the framework together, spacing the studs 24 inches on center.

4. Square the framework and brace it so it will stay square while the sheathing is being nailed.

5. Apply the sheathing, being careful to keep the ends flush with the outside edges of the two end uprights (Figure 9-5).

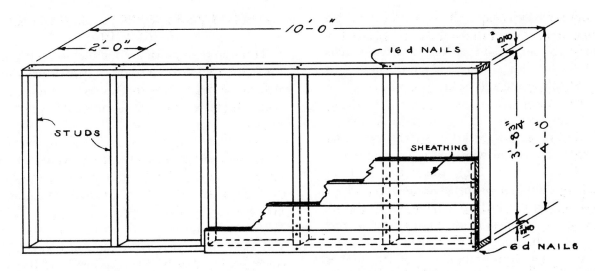

Building a Form Panel
Figure 9-5

HOW TO MAKE COMPLETE WALL FORMS

NOTE: The type of form described here may be used repeatedly in duplicate or near duplicate buildings. It is assumed that forms are to be built for a building 20 feet x 100 feet. All panels are to be 4 feet high.

1. Make twenty panels, as described above, each 10 feet long, for the two outer side walls.

2. Make two panels 10 feet long and two 10 feet 9 inches long. One of each of these is to be used for each outside end wall.

NOTE: These lengths are arrived at as follows: As the end forms for the outside of the wall project over the concrete line of the side forms, 4½ inches must be added to each end panel (A, Figure 9-6). Thus the outside end forms described in Step 2 should be 20 feet plus 4½ inches plus 4½ inches or a total length of 20 feet 9 inches. This can be divided up conveniently into two panels, one 10 feet

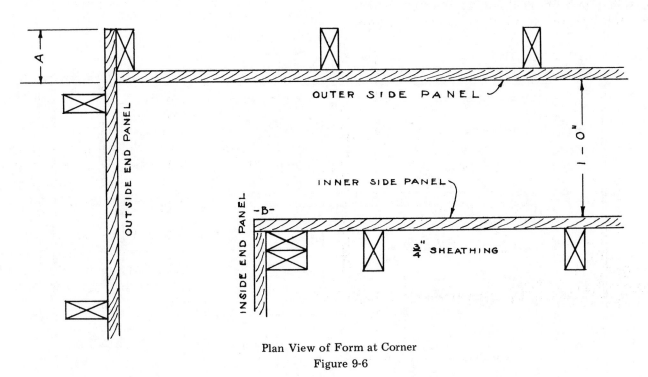

Plan View of Form at Corner
Figure 9-6

83

and one 10 feet 9 inches.

3. Make two panels 10 feet long and two panels 7 feet 10 ½ inches long. One of each of these is to be used for each inside end wall.

NOTE: To arrive at the length of these panels, subtract the combined thickness of the two sidewalls (2 feet) from the width of the building (20 feet), obtaining 18 feet. From this figure, subtract the combined thickness of the sheathing on each of the two sidewalls (1½ inches) (Figure 9-6). The result, 17 feet 10 ¼ inches, is the total length of this form and may be conveniently divided into two panels, one 10 feet long and one 7 feet 10 ½ inches long.

4. Make eighteen panels 10 feet long and two panels 8 feet long for the inner sidewalls. Omit one end stud from each of four of the ten foot panels to allow for nailing the side panel to the end panel. See B, Figure 9-6.

NOTE: The total length of each inner sidewall is 98 feet (the length of the building, 100 feet, minus the combined thickness of the two end walls, 2 feet). Therefore, each side may be made up of nine 10 foot panels and one 8 foot panel.

5. Grease or oil the inside faces of all panels.

HOW TO BRACE AND ALIGN DOUBLE WALL FORMS

1. Set the outside wall panels in place on the footing, bracing them temporarily against the outside of the excavation. The panels may be spiked together (Figure 9-7) or clamped.

2. Provide the necessary holes for the type of spreaders and ties that are used in the panels.

3. Erect the inside panels in a like manner. Double the inside corner studs as shown at B, Figure 9-6. Hold the panels in position by the braces C and D and the stays E, Figures 9-8 and 9-9. Place these braces every 4 feet the length of the wall but nail them only temporarily until after the walers and wall ties have been put in place.

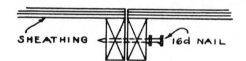

Method of Fastening
Panels Together
Figure 9-7

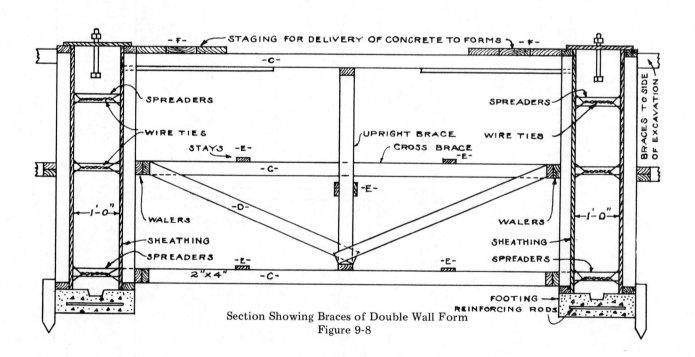

Section Showing Braces of Double Wall Form
Figure 9-8

84

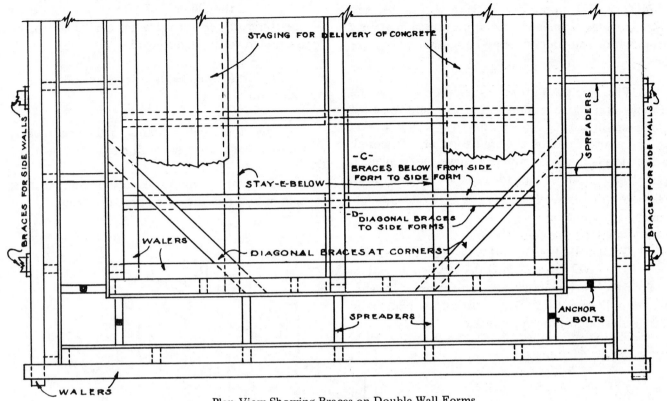

Plan View Showing Braces on Double Wall Forms
Figure 9-9

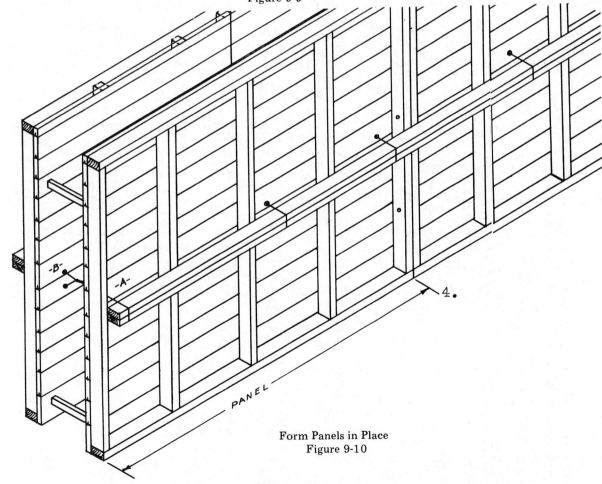

Form Panels in Place
Figure 9-10

85

NOTE: The upper brace (C) in Figures 9-8 and 9-9 provides a means of supporting a platform from which the concrete may be poured.

NOTE: Figure 9-10 shows another view of the double wall with walers, tie wires and wooden spreaders in place. Notice how the panel joints on the opposite sides of the wall form are staggered. After the spreaders and ties have been placed, plumb the double wall form as a complete unit, and then nail the braces rigidly in place.

HOW TO MAKE DOUBLE WALL FORMS WITH CLEATED PANELS

1. Make the panels any convenient length as shown at A, Figure 9-11. The width of the panel may vary according to the type of wall.

2. Space the cleats (B-B) so they come between the studs of the form framework as shown in the assembled form (Figure 9-11).

3. Cut the studs as described in building a single wall form.

4. Nail the cleated panels to the studs. Stagger the joints of the panels on the studs.

5. The remaining procedure is similar to that used in building the other type of forms.

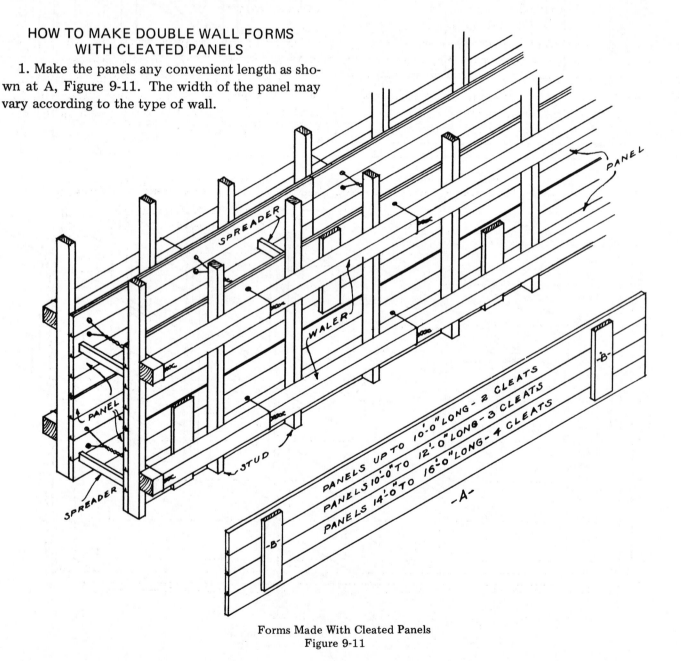

Forms Made With Cleated Panels
Figure 9-11

CHAPTER 10

FORMWORK FOR OPENINGS IN CONCRETE WALLS

In continuous concrete walls, provision must be made in the wooden forms for openings for windows, electric conduits, water pipes, sewer pipes, and bearing seats for girders.

HOW OPENINGS IN CONCRETE WALLS ARE FORMED

Openings are formed by making rectangular box-like forms of wood to the size of the openings to be left in the concrete wall. These box forms are fastened between the wooden forms of the wall so that the concrete, when poured into the forms, will surround the box form, thereby leaving a space in the wall where no concrete can enter. This, in turn leaves a void in the wall the size of the box form (Figure 10-1).

Wood is generally used in making rectangular forms for openings in walls, while sheet metal and the newer type of plywood are used for circular or irregular forms.

Openings for conduits, pipes and other supply lines should be so placed that they will do no structural harm to the concrete wall by weakening it. If it is necessary to place such an opening in a vital part of the wall, the concrete should be reinforced to take up the strain.

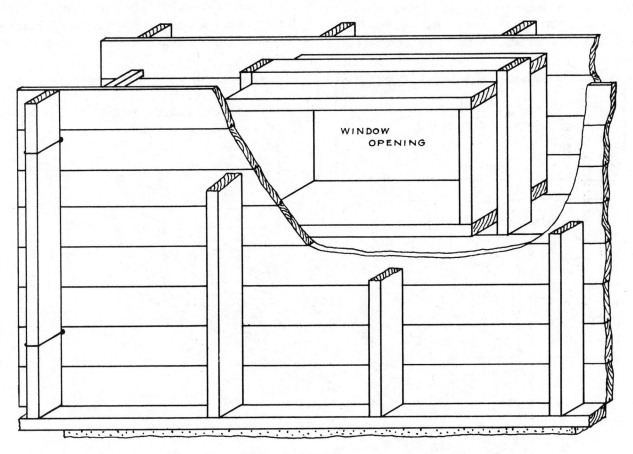

WINDOW OPENING

Form For Opening In Wall
Figure 10-1

These box forms must be constructed in such a manner that they may be removed from the wall after the concrete has hardened. Greasing the surfaces of the forms that are to come in contact with the concrete will make this removal easier by preventing the concrete from sticking to the wood. These forms should be built with a slight taper toward one wall. This makes it possible to slip the box out of the larger end of the hole after the concrete has set. The box-like forms may also be made so that they are collapsible and can be taken apart while still in the hardened concrete wall.

Forms for openings in concrete walls must be so planned and constructed that they will provide for the opening in the proper place and of the correct size. These forms must keep all concrete out of the openings and, in some cases, must withstand considerable pressure from the concrete.

HOW TO MAKE A BOX FORM

NOTE: Assume that it is desired to make a form for an opening for a support running into the wall or a pipe through the wall. The opening is to be 4 inches by 4 inches in a 12 inch concrete wall.

1. Cut two pieces ¾ inch x 4½ inches x 12 inches and taper them as shown in Figure 10-2. These are to be used for the top and bottom of the box form.

2. Cut two pieces 3/4 inch x 3 inches x 12 inches and taper them as shown in Figure 10-3. These are to be used for the two sides of the box form.

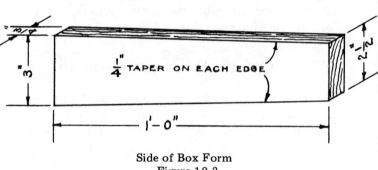

Side of Box Form
Figure 10-3

3. Assemble the four pieces as shown in Figure 10-4 using 6d nails. The assembled box should then measure approximately 4 inches x 4 inches on one end, and 4½ inches x 4½ inches on the other.

4. Round off the outside corners of the box.

5. Cover the outside surface of the box with grease or similar material.

Top and Bottom of Box Form
Figure 10-2

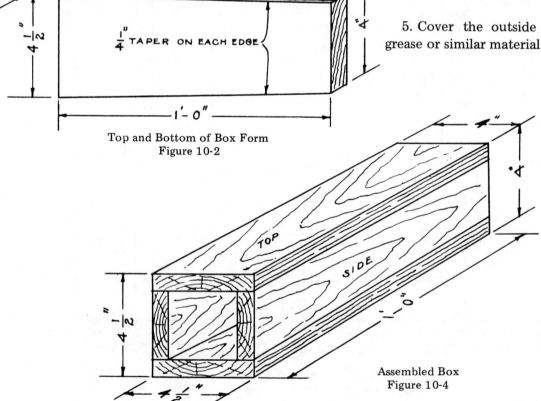

Assembled Box
Figure 10-4

6. Place the box form in the proper position in the concrete forms (Figure 10-5).

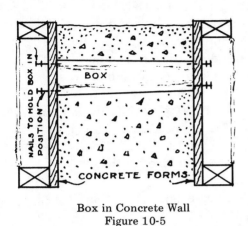

Box in Concrete Wall
Figure 10-5

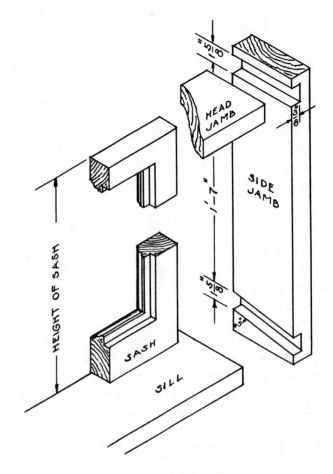

Layout of Sidejamb
Figure 10-6

7. Hold the box in position in the form by driving nails through the sheathing into the ends of the box. Do not drive the nails all the way into the wood, but allow the heads to stick out beyond the surface so they can be pulled out easily when taking the forms off the concrete wall (Figure 10-5).

HOW TO MAKE FRAMES FOR DOOR OR WINDOW OPENINGS

1. Mark the height of the sash (19 inches in this case) on the inside edge of each side jamb (Figure 10-6). These points mark the inside lines of the head and sill dado joints. Allow about 3 inches on each end of the jambs to provide for the dado joints and the horns (Figure 10-7).

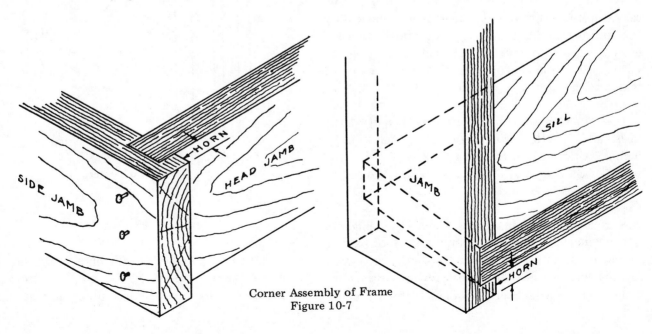

Corner Assembly of Frame
Figure 10-7

2. Lay out the dadoes 1-5/8 inches wide and 3/8 inch deep for the head jamb. These dadoes are square across the boards.

3. Lay out points for dadoes of the same size for the sill, on the inside edges of the side jambs. Be sure to lay out one right and one left jamb.

4. Mark the dado lines across the faces of the boards, using the sliding T bevel set at 5 degrees to 8 degrees. This angle will provide for the pitch of the window frame sill.

5. Cut the four dadoes on the two side jambs.

6. Cut the head jamb and sill 7/8 inch longer than the outside width of the sash. This extra length is added to allow the head jamb and sill to be set into the dadoes which are each 3/8 inch deep. The additional 1/8 inch permits the completed frame opening to be left 1/8 inch oversize since the sash comes considerably larger than nenessary. This saves extra planing in fitting the sash.

7. When the glass size of the window is shown on the plans or is known, the outside size of the sash is figured as follows:

Assume that it is necessary to determine the outside size of a cellar sash of three lights each 12 inches wide and 14 inches high.

A. To find the outside finished width of this sash, multiply the width of the glass (12 inches) by the number of lights (3). To this add ¼ inch for each of the two muntin bars between the glass. Also, add 2 inches for each side rail. Thus the width of the sash is 40½ inches.

B. To find the outside finished height of the sash, add together the height of the glass (14 inches), the width of the top rail (2 inches) and the width of the bottom rail (3 inches). Thus the height of the sash is 19 inches.

8. If 1-5/8 inch side jambs are used, casings are not needed. If 3/4 inch jambs are used, casings should be provided and the jambs should be braced so they will not bow from the pressure of the concrete. Jambs which are 1-5/8 inch thick are the

ones most commonly used for cellar frames.

9. When provision is to be made for door openings in concrete walls, the procedure is similar to that used for window openings. Care must be taken, however, to see that the side jambs or bucks of the door frames are braced so the concrete will not bow them out of plumb.

HOW TO ASSEMBLE THE FRAME

1. Set the head jamb and sill into the dadoes in the side jambs. Nail through the side jambs into the head jamb and sill (Figure 10-7).

2. Square and brace the frame as shown at A, Figure 10-8.

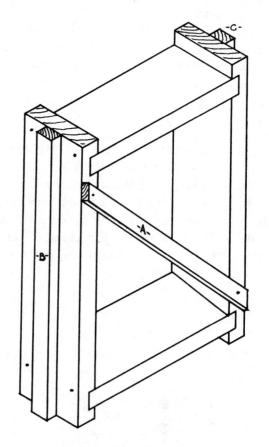

Assembly of Frame
Figure 10-8

3. Cut off the horns at the ends of the side jambs if necessary (Figure 10-7). The bottom horns should be cut square across rather than at an angle.

4. Nail strips to the outside of the side jambs and sill to hold the frame in the concrete (B and C, Figure 10-8).

TO INSTALL THE FRAME IN THE FORM

Nail the frames in position in the forms as in Figure 10-1. If the side jambs of the frame are narrower than the inside width of the form, the frame may be nailed to the inside or outside form as desired, and pieces of 2 x 4 may be nailed in the gaps between the frame and the form. This will prevent concrete from running into the frame opening. See Figure 10-9.

NOTE: When a finished window or door frame is to be set in a concrete wall after the wall has been poured, a temporary frame or buck is placed in the form. Bevel shaped key strips should be tacked on the outside of the buck. These strips remain in the hardened concrete after the buck has been removed from the wall, to provide a nailing surface for the finished frame.

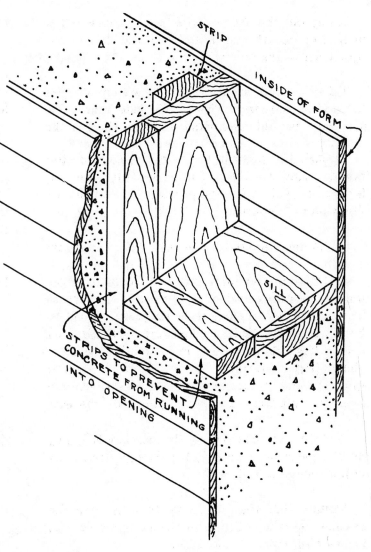

Cellar Frame in Concrete Form
Figure 10-9

CHAPTER 11

FORMWORK FOR STEPS

Concrete steps are superior to wooden steps for outside approaches. This is especially true where the grade level changes, such as in terraces, or where the steps are embedded in the ground. This type of step is also used in outside hatchways to cellars where the stairway is subjected to hard use.

Concrete steps will not rot, they are more permanent than wooden steps and are easier to keep clean. They can be provided with safety treads by sprinkling powdered carborundum on the treads just before the concrete sets. Concrete steps can also be made waterproof by a coating painted over the surfaces. Iron hand rails and nosings can be installed solidly in the concrete treads or sidewalls to provide a safe stairway.

TYPES OF CONCRETE STEPS

There are two general types of concrete steps. One type is built directly on the slope of the ground, while the other is supported at the top and the bottom, leaving an air space under the steps.

Steps Built On Sloping Ground. This type of step is sometimes built between concrete walls. The wooden forms for such a wall are constructed in much the same manner as those of a regular continuous wall.

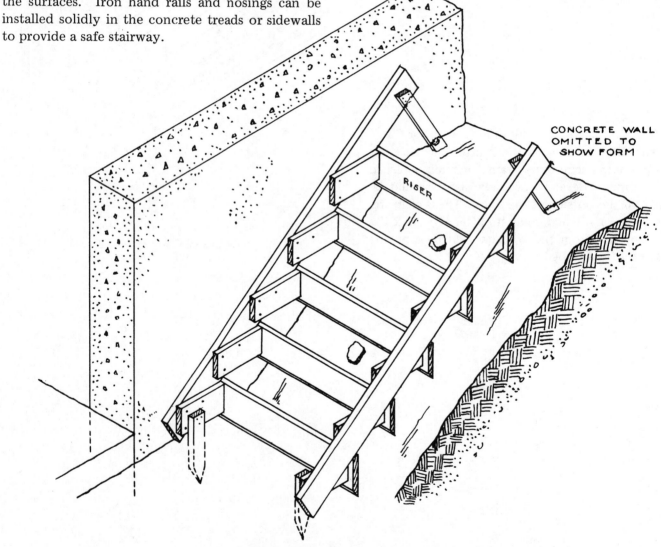

CONCRETE WALL
OMITTED TO
SHOW FORM

RISER

Forms for Steps Built on Sloping Ground
Figure 11-1

92

After the concrete in the sidewalls is hard enough to permit removal of the forms, the inside form of each sidewall is carefully removed. The forms that are to be used to form the risers and treads of the steps are then built between the sidewalls (Figure 11-1).

Blocks are nailed to the hanging stringers (Figure 11-1) along the layout marks for each tread and riser. The riser boards are then nailed to these blocks. The slope of the ground is often stepped off so the concrete will not slide along the bed of the steps.

Another example of steps built on sloping ground is shown in Figure 11-2.

The soil must be well drained under this type of step so water cannot accumulate underneath them. If the soil is clay, a well tamped layer of cinders should be placed underneath the 4 inch concrete bed of the steps.

Figure 11-2 shows a completed concrete stairway for an outside entrance to a cellar.

The Open Type Of Steps. The forms for the open type of concrete steps (Figure 11-3) are somewhat more difficult to build.

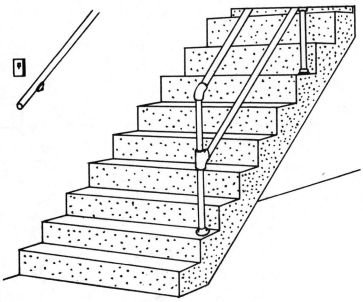

Open Concrete Steps
Figure 11-3

A wooden platform of tongue and groove sheathing is built to support the bottom or supporting slab of the stairs. See Figure 11-4. This panel extends about 12 inches beyond each side of the stairs to provide a support for stringer bracing blocks. The back of the panel should be well braced and supported by 4 x 4's. These should rest on wedges to permit easy adjustment and to simplify the removal of the posts.

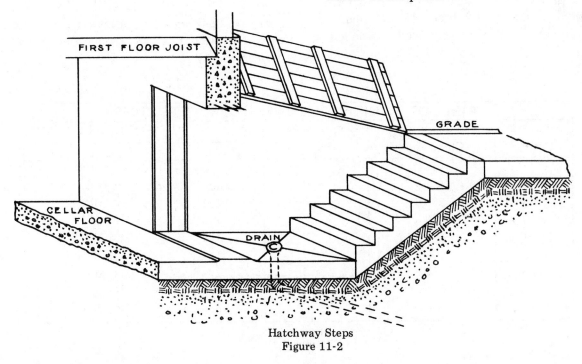

Hatchway Steps
Figure 11-2

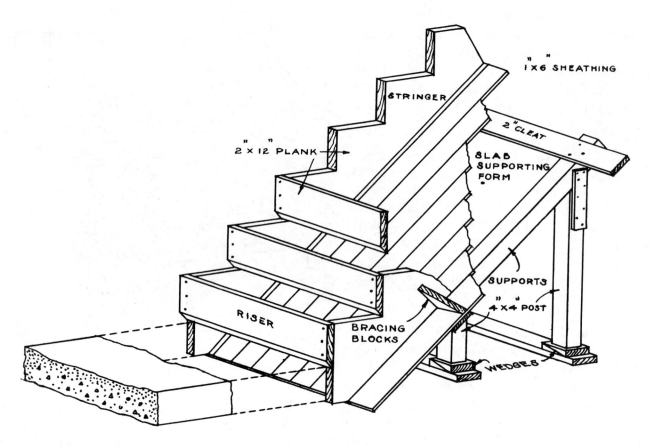

Forms for Open Concrete Steps
Figure 11-4

The side stringers are laid out in the same manner as the supporting carriage for a flight of wooden stairs and are cut from a 2 inch x 12 inch plank. The pieces that are cut out may be saved and nailed to a piece of 2 x 4 or 2 x 6 to make an additional stringer if needed.

Figure 11-5 shows another method of building forms for steps of long span, or subject to heavy service.

TABLE OF REINFORCEMENT FOR CONCRETE STEPS				
No. Of Steps	Clear Span	Thickness Slab	REINFORCEMENT	
			Diameter	Spacing Rods
4	2'- 2"	4"	1/4"	10"
5	3'- 0"	4"	1/4"	10"
6	3'-10"	4"	1/4"	7"
7	4'- 8"	5"	1/4"	7"
8	5'- 6"	5"	1/4"	5"
9	6'- 4"	6"	1/4"	5"
10	7'- 2"	6"	3/8"	5"
11	8'- 0"	6"	3/8"	6"

Table 11-1

94

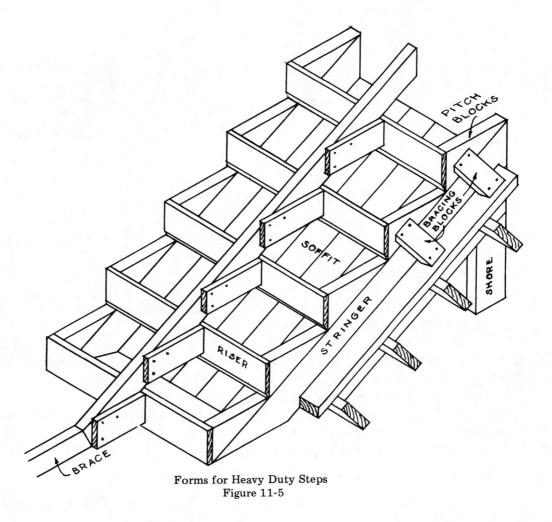

Forms for Heavy Duty Steps
Figure 11-5

Reinforcing rods should be placed 1 inch from the surface of the form supporting the slab. The rods should run lengthwise up the platform and should be spaced as shown in Table 11-1. The location of these rods in the finished steps is shown in Figure 11-6.

It is customary to remove the side forms and riser forms about 24 hours after the concrete has been poured, so the troweled finishing coat of cement will bond with the faces of the treads, risers and side walls. The supporting forms, such as the platform and the shoring, should remain intact about three weeks.

STAIR RISER FORMS

There are several different types of forms for risers of concrete stairs. The most common type (Figure 11-7, A) is made from a straight plain board and forms a plain faced concrete riser. At B the same type of board is used, but it is tilted so as to

form an undercut concrete riser face. At C the riser board is built up of two pieces so as to give the concrete riser face an ornamental or panel effect.

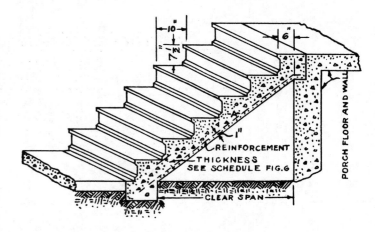

Section Of Steps Showing Reinforcing Rods
Figure 11-6

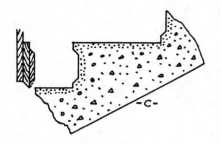
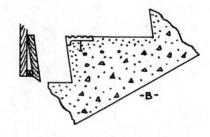
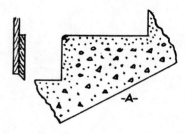

Detail Of Step Treads And Forms
Figure 11-7

The bottom edge of all the riser boards is beveled to allow the mason to trowel the face of the tread back to the face of the riser while the forms for the risers are still in place.

HOW TO FIND THE RISE AND RUN OF THE STAIRS

Before the forms can be put in place the stairs must be designed. Note the side wall section depicted in Figure 11-8. The point A represents the head of the steps and the upper sidewalk level. The point B represents the starting point of the steps and the lower sidewalk level. The distance

A-C represents the run or horizontal length of the stairs.

1. Extend a level line on the side wall from point A as shown in Figure 11-8.

2. From point B, mark a plumb line on the wall, extending it so it intersects the level line from point A. This establishes point C.

3. Measure the vertical distance between points B and C. In this case it is 42 inches. This represents the total rise of the stairs.

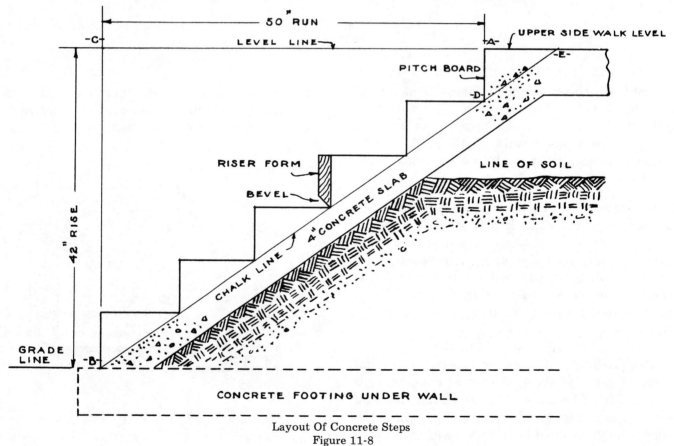

Layout Of Concrete Steps
Figure 11-8

4. To find the riser height, divide 42 inches, the total rise of the stairs, by 7 inches, the allowable height of each riser for outside stairs. The result, six, is the number of risers required.

NOTE: If the total rise of the stairs is in inches and fractions of an inch, lay the total rise off in equal divisions by using dividers. This will give the height of each riser and the number of risers required. Care should be taken to make the divisions 7 inches or less.

5. Determine the width of the tread.

NOTE: A good method of proportioning risers and treads is to add to any given height of riser that number which will make a total of twelve. Double the number added, and the result will be the width of the tread.

To find the width of the tread in this case, add to the rise that number which will make a total of twelve (7 + 5 = 12). Double the number added to the rise (5 x 2 = 10) and the result, 10, will be the required width of the tread.

If the number of risers and the total run of the stairs are definitely fixed, the width of the tread may be found by dividing the run (50 inches) by the number of risers (5). Thus the width of the tread will be 10 inches.

The number of treads required is always one less than the number of risers.

HOW TO MAKE AND USE A PITCH BOARD

1. To make a pitch board (Figure 11-9), lay the steel square on a piece of 1 inch x 8 inch stock so the figures 7 on the tongue and 10 on the body of the square coincide with the edge of the board as at D and E, Figure 11-10.

2. Mark the lines AD and AE on the board along the tongue and body of the square.

3. Cut on these lines and the result will be a pitch board as shown in Figure 11-9.

4. Plumb down from point A, Figure 11-8 a distance equal to the height of a riser. This locates

point D.

5. To mark the pitch of the steps, connect points B and D, Figure 11-8 by snapping a chalked line on the face of the concrete side wall. Extend this line from D so it meets the upper sidewalk level at E.

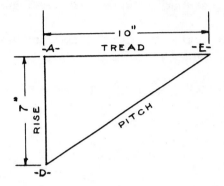

Figure 11-9

6. Place the pitch board on the concrete side wall in such a position that the point E, Figure 11-9, lies directly on point E, Figure 11-8 and that the line DE, Figure 11-9, coincides with the line EDB, Figure 11-8.

7. Mark the wall by scribing along the outside edges of the pitch board. The 10 inch edge represents the tread of the steps and the 7 inch edge represents the rise.

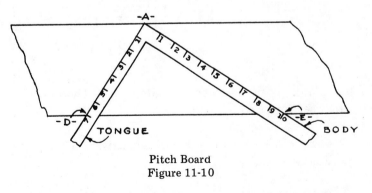

Pitch Board
Figure 11-10

8. Repeat this procedure down the line EDB, Figure 11-8 until all the steps have been marked.

9. Repeat steps 4, 5, 6, 7 and 8 on the opposite concrete side wall.

10. Check the tread marks on each wall to see that they are level with each other. If so, the riser

marks will also check.

HOW TO INSTALL THE STRINGERS

1. Put 2 inch x 12 inch planks in position, so the under side of the plank just clears the nosing of each step as marked on the wall. See Figure 11-11.

2. Cut in the two 2 x 4 braces and wedge the stringers tightly in position against the side walls.

3. Cut six pieces of 1-5/8 inch x 7 inch dress-

NOTE: A cut-out stringer, such as shown in Figure 11-12 may also be used for this type of stairs.

6. Nail the riser boards to the cleats, keeping the top and bottom edges of the riser boards even with the tread marks on the side wall.

7. Brace the stringers at the bottom so the weight of the concrete will not force them down (Figure 11-11).

8. Check the tops of the riser boards to see if

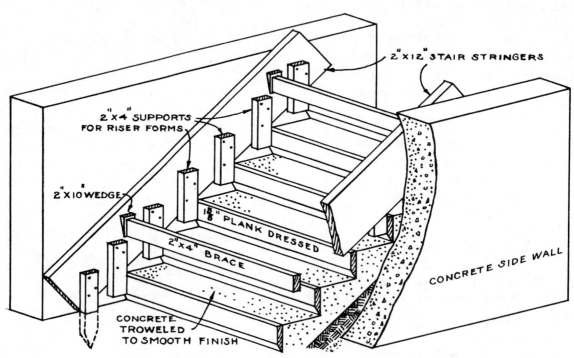

Forms For Steps With Concrete Side Walls
Figure 11-11

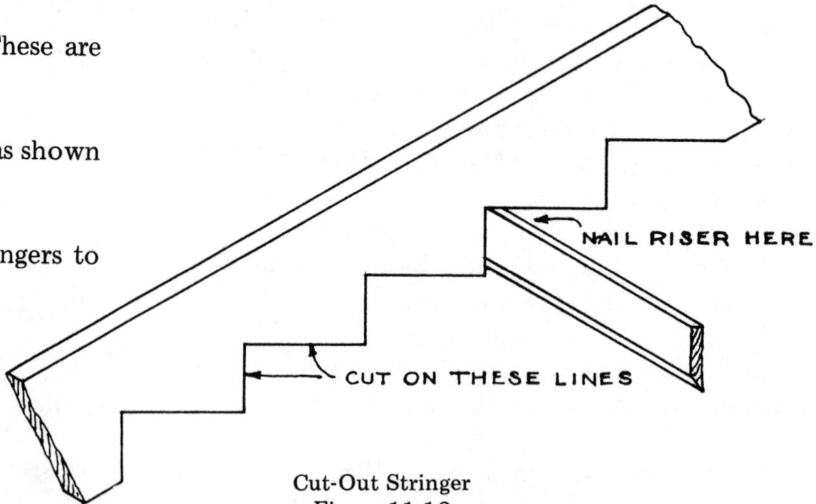

ed plank to fit between the side walls. These are the riser forms.

4. Bevel the bottom edge of each riser as shown in Figure 11-11.

5. Nail short pieces of 2 x 4 to the stringers to provide a surface on which to nail the ends of the riser boards (Figure 11-11). Keep the 2 x 4 cleats back the thickness of the riser boards from the riser marks on the side wall. The bottom end of the cleats should be about one inch above the tread mark on the side wall.

Cut-Out Stringer
Figure 11-12

98

they are level. Recheck all members of the form to be sure they have not moved, or will not move from the pressure of the concrete.

NOTE: Figure 11-11 shows the forms in place between side walls. The forms for the side walls are made of panel sections similar to those described in Chapter 8. The aligning, bracing, and spacing is also the same as in the construction of forms for a continuous concrete wall.

HOW TO BUILD OPEN STEP FORMS

1. Lay out line B, Figure 11-13, for the bottom of the sheathing.

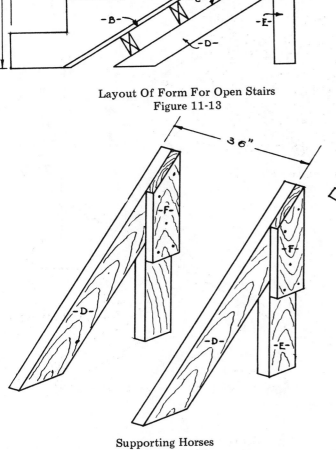

Layout Of Form For Open Stairs
Figure 11-13

Supporting Horses
Figure 11-14

2. Lay out line C. This is the top of the inclined member D. Figures 11-13 and 11-14. From this line, the length and bevels of the inclined piece D can be obtained.

3. Cut two pieces D, and try them in place on the full size layout to see if they are correct.

4. Determine the length and bevel of the shore E, Figures 11-13 and 11-14, and cut two of these.

5. Nail the pieces D and E together with cleats F as shown in Figure 11-14.

6. Place the assembled horses squarely against the walls where the form is to be located.

7. Nail cross joists (Figure 11-15) to the inclined tops of the horses with only enough nails to hold them in place. These pieces should project at least 14 inches on each side and should be spaced about 16 inches on centers.

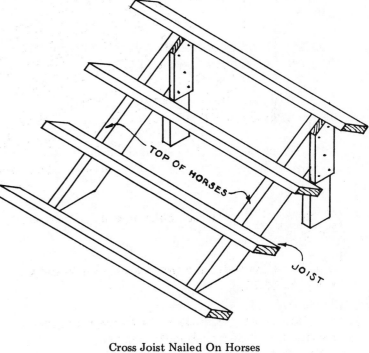

Cross Joist Nailed On Horses
Figure 11-15

8. Nail the sheathing in place to form the bottom of the stair soffit (Figure 11-16). Lay the sheathing from each side to the center and put in the fill-

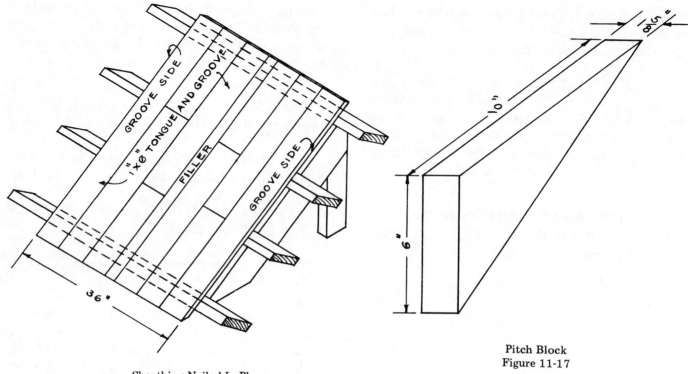

Sheathing Nailed In Place
Figure 11-16

Pitch Block
Figure 11-17

er piece last. Short pieces can be used to advantage as shown.

9. Lay out and cut the pitch blocks (Figure 11-17). Refer to Figures 11-8 and 11-9.

10. Nail the pitch blocks to a plank to form the stringer (Figure 11-8).

11. Place the built up stringer on each side of the form as shown in Figure 11-18.

12. Nail 2 x 8 brace supports on both sides of the stairs as shown in Figure 11-18. Do not drive these nails home.

13. Plumb the stringers of the stairs and put in braces from the brace supports to near the top of the pitch blocks on the built up stringer.

14. Cut the riser boards to length and width, and bevel them at the bottom edge. The bottom riser board need not be beveled.

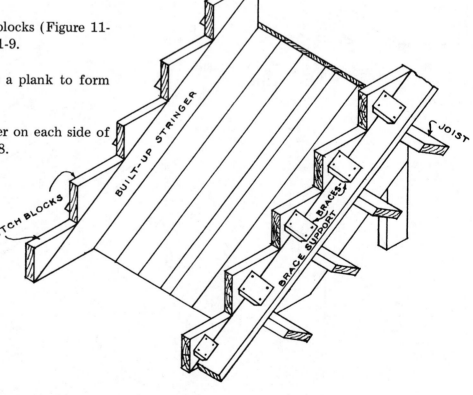

Stringers In Place
Figure 11-18

15. Nail on the risers as shown in Figure 11-19.

16. Check the tops of the riser boards and the tread line of the pitch blocks to see if all are level. Secure the braces as explained for step forms built on sloping ground.

17. Place reinforcement rods 1 inch from the surface of the sheathing as explained in Table 11-1 and Figure 11-6.

Figure 11-20 shows the completed open type of concrete stairs. The rise of each step is 6 inches and the total rise is 30 inches. The treads are 12 inches and the total run is 48 inches from the face of the platform.

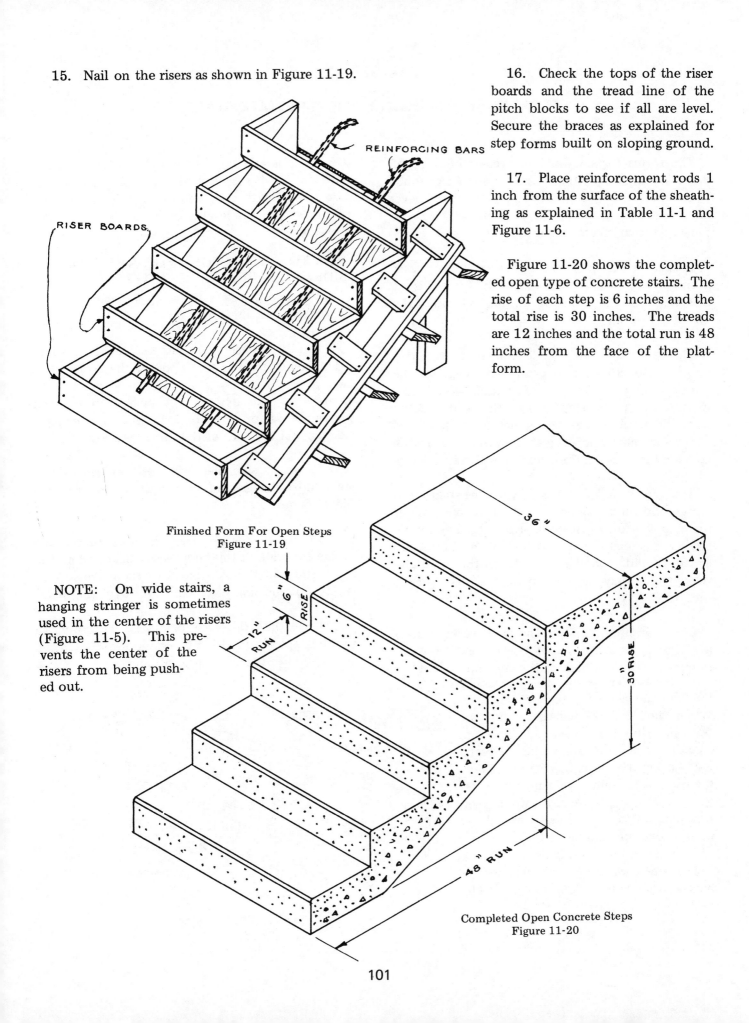

RISER BOARDS

REINFORCING BARS

Finished Form For Open Steps
Figure 11-19

NOTE: On wide stairs, a hanging stringer is sometimes used in the center of the risers (Figure 11-5). This prevents the center of the risers from being pushed out.

36"

6" RISE

12" RUN

30" RISE

48" RUN

Completed Open Concrete Steps
Figure 11-20

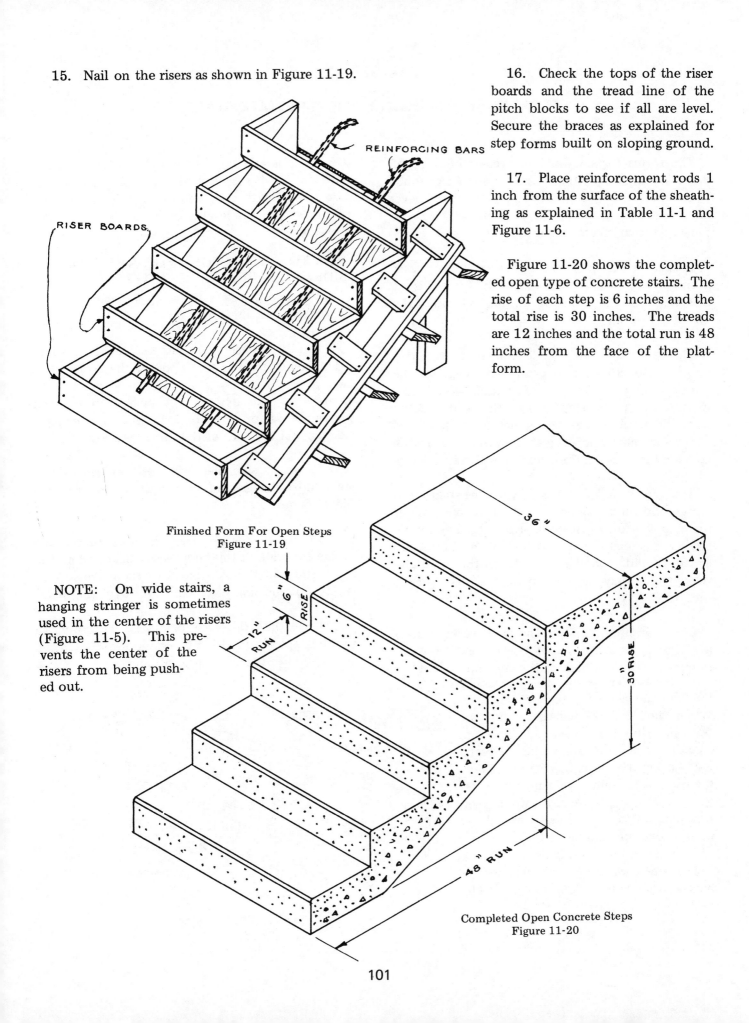

CHAPTER 12

FORMWORK FOR FLOOR AND SIDEWALK SLABS

The forms for sidewalks are generally made of 2 x 4 stock. Only straight and sound stock should be selected. Long lengths should be used whenever possible, as the fewer joints there are in the form, the easier it will be to straighten it.

Forms for sidewalks and floors should be placed in such a way that the top edge of the form will act as a guide for the mason in finishing the top of the sidewalk or floor.

The forms should be made straight and true and properly sloped so as to provide drainage for the finished concrete surface. For drainage of sidewalk surfaces, the slope across the width of the sidewalk should not exceed ¼ inch for each foot of width. In cellar floors the amount of slope may be increased so that the floor surface will drain readily.

This type of form is braced by driving stakes into the ground on the outside of the form and nailing them to the form. Additional bracing may be necessary, especially if the bottom of the form does not rest solidly on the ground. In this case, gravel or earth may be used to form a solid bed for the form.

Concrete sidewalk slabs should have expansion joints to provide for the expansion and contraction caused by the heat of the sun and the cold of the winter. These joints are made by inserting a tapered piece of wood between the concrete panels as they are poured. They are placed so as to act as a guide for the concrete workers in forming the line between the panels of concrete when finishing the top coat. After the finished concrete has partially set, the piece of tapered wood is pulled out of the concrete, and melted tar or a plastic caulking compound is run in the slot. This forms a water tight joint as well as an expansion joint for the concrete.

In making and setting forms for concrete walks and floors, it is essential that they be placed at the correct height so they will drain well and so they will meet ground levels, floors, steps, and walks properly.

HOW TO BUILD A FORM FOR A SECTION OF SIDEWALK 4 FEET WIDE AND 12 FEET LONG

1. Cut two pieces 2 inches by 4 inches x 12 feet for the sides of the form.

2. Cut two pieces 2 inches x 4 inches x 4 feet 3-1/4 inches for the two ends of the form. Cut two pieces 2 inches x 4 inches x 4 feet for the spreaders. The extra 3-1/4 inches on the end pieces allows for a lap of 1-5/8 inches over each side piece at the ends of the form. See Figure 12-1.

3. Place the side pieces in position and drive a stake into the ground near one end of the side piece (Figure 12-1).

4. Determine the required elevation of the top of the sidewalk at this end. Mark this point on the stake and nail the side piece to the stake so that the top of the side piece coincides with the mark.

5. Level the side piece, or pitch it according to the drainage required for the sidewalk. Fasten the second end of the side piece to a stake.

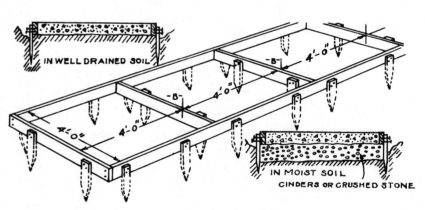

Form For Concrete Sidewalk
Figure 12-1

6. Stretch a chalk line from one end stake to the other and align the edge of the side piece with the chalk line.

7. Drive stakes every 3 feet along the side piece and nail them to it (Figure 12-1).

8. Repeat the same procedure for the other side of the form.

9. To prevent the sides of the form from being crowded together, place the spreader pieces where the joints in the panels are to occur. These spreaders should remain between the sides until the coarse layer of concrete has been placed in the form.

10. Nail the end pieces of the form in position, keeping the top edges even with those of the sides (Figure 12-1).

11. Test the top and side edges of the com-pleted form with a straight edge and make any necessary adjustments.

12. Pack earth around the outside and bottom edges of the form to keep it from moving if it is stepped on.

HOW TO MAKE CURVED SIDEWALK FORMS

1. Lay out the required arc or curve to full size. The subfloor of the building is a convenient place for this layout work.

2. If the arc or curve is large, toenail 2 x 4 blocks on edge, about 1 foot apart along the curved line on the subfloor (Figure 12-2).

3. Stand two pieces of ½ inch x 4 inch stock on edge on the floor and between the starting blocks shown in Figure 12-2. These two boards should be longer than the curve. Nail the blocks solidly to the subfloor.

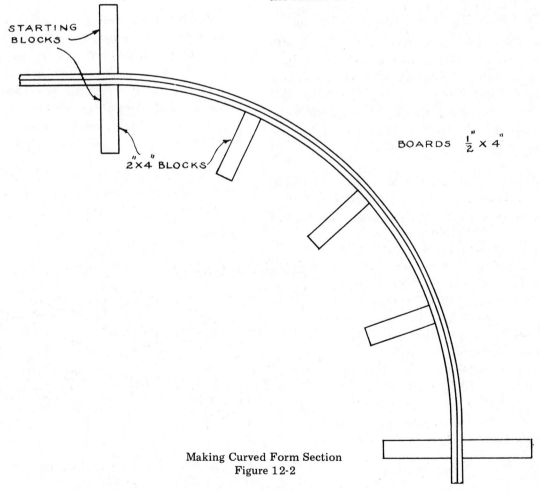

STARTING BLOCKS

2"X4" BLOCKS

BOARDS $\frac{1}{2}" \times 4"$

Making Curved Form Section
Figure 12-2

4. Gradually bend the boards against the blocks, keeping the face of the boards vertical and sliding them over the floor. When the end of the curve is reached, nail a block on the floor to hold the boards in this position.

5. Nail the two boards together about every 6 inches with shingle nails. If the nails come through the two boards, clinch them over.

NOTE: If the boards do not bend readily, soak them in water for about one hour. Basswood or white pine will bend under these conditions.

6. After the boards have been nailed, place a cross brace on them. The blocks may be removed and the bent member placed in position in the form.

NOTE: If the curve is small, the piece may be cut from a solid piece of wood, or sheet metal or plywood may be used.

HOW TO PROVIDE FOR EXPANSION JOINTS

1. Cut pieces of 5-3/4 inch clapboards 4 feet long.

2. Drive shingle nails 2 inches from the butt edge of the clapboards and hang them on the top

of the spreaders to hold them at the proper height (Figure 12-3).

NOTE: Sometimes special expansion joints are furnished by the mason.

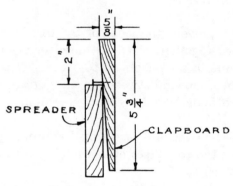

Expansion Joint Section
Figure 12-3

NOTE: The forms for interior concrete floors such as cellar floors, are generally made in a similar manner as those for sidewalks, except that the expansion joints and panels are omitted. Straight edges are used to establish the top of the forms at the drainage points.

CHAPTER 13

HOW TO MAKE BEAM AND GIRDER FORMS

Beam Form. The type of construction to be used for beam forms depends upon whether the form is to be removed in one piece or whether the sides are to be stripped and the bottom left in place until such time as the concrete has developed enough strength to permit removal of the shoring. The latter type beam form is preferred; details for this type are shown in Figure 13-1. Beam

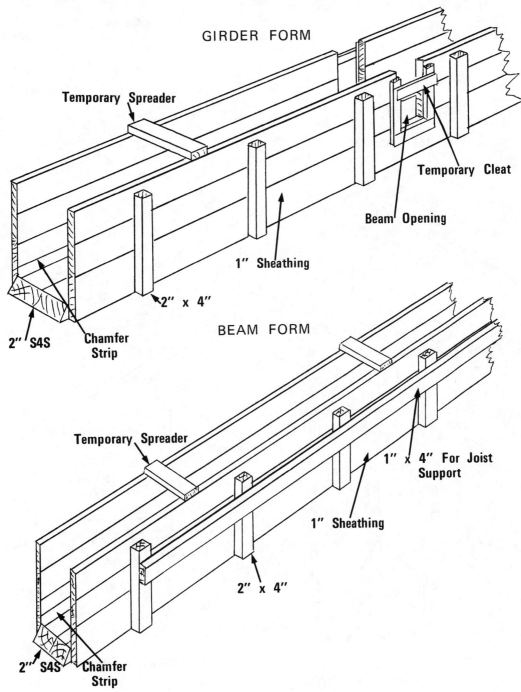

GIRDER FORM

Temporary Spreader

Temporary Cleat

Beam Opening

1" Sheathing

2" x 4"

2" S4S

Chamfer Strip

BEAM FORM

Temporary Spreader

1" x 4" For Joist Support

1" Sheathing

2" x 4"

2" S4S

Chamfer Strip

Beam and Girder Forms
Figure 13-1

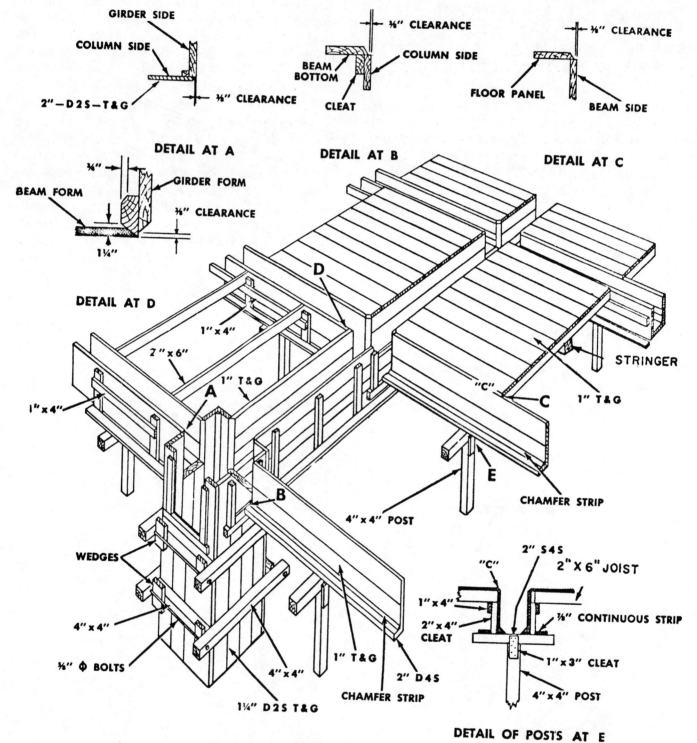

GIRDER SIDE

COLUMN SIDE

2"—D2S—T&G

¾" CLEARANCE

DETAIL AT A

⅛" CLEARANCE

BEAM
BOTTOM

COLUMN SIDE

CLEAT

DETAIL AT B

⅛" CLEARANCE

FLOOR PANEL

BEAM SIDE

DETAIL AT C

¾"

BEAM FORM

GIRDER FORM

⅛" CLEARANCE

1¼"

DETAIL AT D

1" x 4"

2" x 6"

1" T&G

1" x 4"

A

WEDGES

4" x 4"

½" Φ BOLTS

1¼" D2S T&G

B

4" x 4"

CHAMFER STRIP

1" T&G

2" D4S

D

4" x 4" POST

"C"

C

STRINGER

1" T&G

E

CHAMFER STRIP

2" S4S

"C"

2" X 6" JOIST

1" x 4"

2" x 4"
CLEAT

⅛" CONTINUOUS STRIP

1" x 3" CLEAT

4" x 4" POST

DETAIL OF POSTS AT E

Assembly Details, Beam and Floor Forms
Figure 13-2

106

forms are subjected to very little bursting pressure but must be shored up at frequent intervals to prevent sagging under the weight of the fresh concrete.

Beam Form Constructions. The bottom of the form has the same width as the beam and is in one piece the full width. The sides of the form should be 1-inch thick tongue and groove sheathing and should lap over the bottom as shown in Figure 13-2. The sheathing is nailed to 2- by 4-inch struts placed on 3-foot centers. A 1- by 4-inch piece is nailed along the struts. These pieces support the joist for the floor panel, as shown. The sides of the form are not nailed to the bottom but are held in position by continuous strips as shown in detail E.

The cross pieces nailed on top serve as spreaders. After erection, the slab panel joints hold the beam sides in position. Girder forms are the same as beam forms except that the sides are notched to receive the beam forms. Temporary cleats should be nailed across the beam opening when the girder form is being handled.

Assembly. The method of assembling beam and girder forms is illustrated in Figure 13-2. The connection of the beam and girder is illustrated in Detail D. The beam bottom butts up tightly against the side of the girder form and rests on a 2- by 4-inch cleat nailed to the girder side. Detail C shows the joint between beam and slab panel and details A and B show the joint between girder and column. The clearances given in these details are needed for stripping and also to allow for movement that will occur due to the weight of the fresh concrete. The 4 by 4 posts used for shoring the beams and girders should be spaced so as to provide support for the concrete and forms and wedged at either the bottom or top for easy dismantling.

CHAPTER 14

FORMS FOR ARCHED OPENINGS

Arches of concrete, brick, or stone are used for door openings, cellar window areas and fireplaces, if it is necessary to support masonry above these openings. The masonry materials of which the arch is composed must be supported during the process of construction. Wooden forms called arch centers are built for this purpose by the carpenter. The arch center also acts as a guide to the mason. The centers may be removed after the mortar is thoroughly dry and the arch has become self-supporting.

For a masonry wall, the arch center is made as shown in Figure 14-2. The lagging, generally consisting of ¾ inch x 2 inch strips of wood, acts as a support for the masonry of the arch and as a means of holding the ribs together. The lagging is cut one inch shorter than the width of the masonry wall. This prevents the ribs of the arch center from interfering with the mason's lines. The ribs form the sides of the arch center and are cut to the shape of the arch.

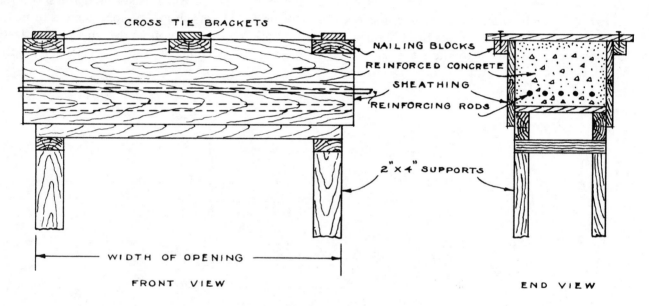

Form For Reinforced Concrete Arch
Figure 14-1

Although there are many types and sizes of arch centers, only a few of the most common ones will be covered in this unit. The layout of the more complex types should be done by the architect.

The flat arch. Figure 14-1 shows an arch center made for reinforced concrete. It consists of a box-like form supported over the width of the opening. The concrete is cast in this form.

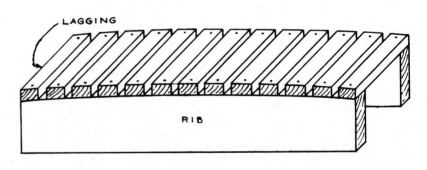

Flat Or Jack Arch
Figure 14-2

Usually this type of arch has a slight crown or rise in the center so the arch will appear straight. If the arch is actually made straight, it will appear to be sagging in the center.

Figure 14-3 shows the finished flat arch faced with a soldier course of brick. The brick is supported by an angle iron spiked on the outer rib of the arch center.

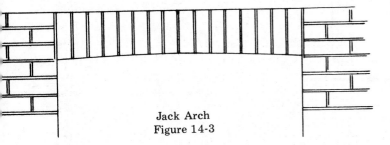

Jack Arch
Figure 14-3

HOW TO BUILD THE FLAT OR JACK ARCH CENTER

NOTE: Assume that an arch center is to be built for a door opening 3 feet 6 inches wide. The masonry wall is to be 1 foot thick and the arch is to have a 1 inch rise.

2. Measure in from each end 20½ inches and square a line across the width of the board. This is the center line as shown at C, Figure 14-4.

3. Measure down from the top edge 1-3/8 inches at each end and drive a nail about half way in at these points.

4. On the center line (C, Figure 14-4) measure down 3/4 inch from the top edge and drive a nail partly in at this point.

5. Procure a strip of wood 3/8 inch thick and about 4 feet long.

6. Place and bend the strip of wood so it is against the three nails as shown in Figure 14-4.

7. Mark a line along the strip, thus outlining the curve.

8. Lay out the other rib in the same manner and saw along the curved lines with a rip saw.

9. Square the ends of a ¾ inch board to 11 inches

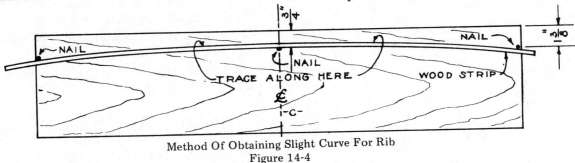

Method Of Obtaining Slight Curve For Rib
Figure 14-4

1. Cut two pieces of stock 2 inches x 6 inches x 3 feet 5 inches. Square both ends of each piece.

long. Rip 2 inch wide lagging strips from this board.

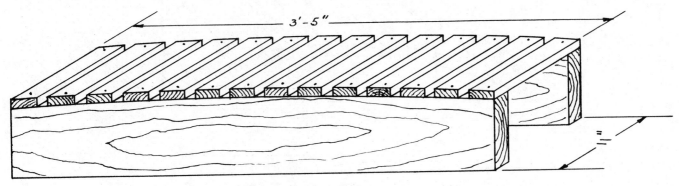

Assembled Arch Center
Figure 14-5

10. Nail the lagging so the ends come flush with the outside face of each rib (Figure 14-5). Space the pieces of lagging about 1 inch apart. Be sure the center is squared, and that a few of the pieces of lagging are double nailed at the ends so as to hold the center square.

NOTE: For the reinforced concrete flat arch, a box-like form is built to hold the wet concrete, as shown in Figure 14-1.

The Segmental Arch. The segmental arch is so called because it forms a segment of a circle. Figure 14-6 shows the layout of this arch. A rib of this type of arch center (Figure 14-7) is made from a wide board, the top of which is cut to the desired curve. Lagging may or may not be used, depending on the thickness of the masonry. If brick or stone veneer is to be supported, the two ribs of the arch center may be nailed to blocks which spread the

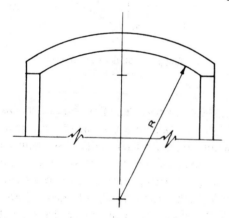

Segmental Arch Layout
Figure 14-6

ribs to within 1 inch of the thickness of the veneer. Figure 14-8 shows the finished arch after the center has been removed.

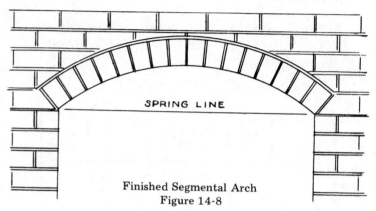

SPRING LINE

Finished Segmental Arch
Figure 14-8

It is advisable to lay out arch outlines by geometrical construction on a large layout board or on the subfloor of the building. The carpenter should provide himself with a set of trammel points, and make a wooden beam on which to mount the points. The beam should be made long enough to mark the radius of the largest arc encountered. Using the trammel points is a more accurate method than using a pencil and line. It is also more accurate than using a wooden rod with nails driven in each end, the distance between the nails representing the radius of the arc.

HOW TO BUILD A SEGMENTAL ARCH CENTER

Assume that the opening in a 12 inch wall is 24 inches wide and the rise of the arch center is to be 6 inches.

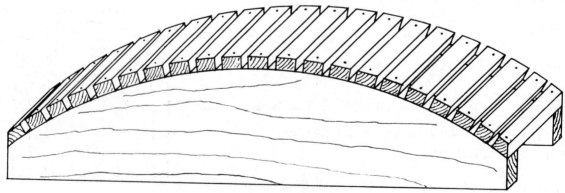

Rib For Segmental Arch Center
Figure 14-7

1. Cut two pieces of stock 1 inch x 10 inches x 24 inches.

2. Square a line across the center of one board.

3. Tack the board on the subfloor and, with the tongue of the framing square against an edge of the board and on the center line, project a line away from the board about 18 inches long (Figure 14-9).

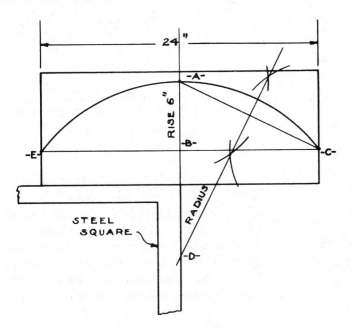

Layout Of Segmental Arch
Figure 14-9

4. From a point (A) on the center line near the top of the board, measure down 6 inches along the center line to establish point B, Figure 14-9.

5. From line AB, square a line across the board to locate point C.

6. Draw the line AC.

7. Construct a bisecting line, vertical to line AC, and extend it until it intersects the center line of the board at D.

8. Using AD as a radius, swing an arc EAC by means of the trammel points.

NOTE: If lagging is to be used, the thickness of the lagging should be deducted from the length AD, thus making an arc with a smaller radius.

9. With a compass saw, cut on the line just marked.

10. Mark and cut the other piece.

11. Cut and nail on the lagging strips as explained for the jack arch center.

NOTE: Bevel the end pieces of lagging slightly so the top edge will not project beyond the ends of the rib (Figure 14-10). The end pieces of lagging may also be kept back so the outside edge does not project beyond the end of the ribs.

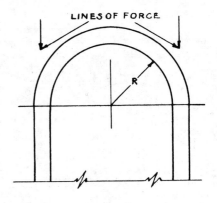

Lagging At End Of Rib
Figure 14-10

The Semi-Circular Arch. The semi-circular arch (Figure 14-11) is perhaps the most common type. In a crowned arch of this type, the weight above the arch is diverted to the sides of the opening which act as supporting piers. The force on these side walls is both down and outward. Therefore, the side walls of crowned arch openings must have solid side support.

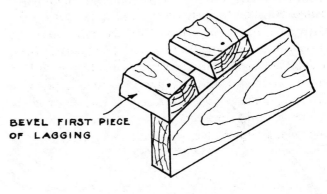

Layout Of Semi-Circular Arch
Figure 14-11

111

Since some of the semi-circular arch centers require many pieces of lumber to make the segmental parts, it is more difficult to lay out this kind of an arch than to lay out one that can be made from a solid board.

The outside edges of the segments A, B, C, Figure 14-12 are cut to coincide with the circumference of the circle. The miter cut (D) is made on a radius of the circle. Cleats are fastened behind these three pieces to hold them together, and ties are used at the bottom to prevent the center from from spreading. In large arches, struts extending from the tie to the segments are used to help carry the load of the masonry. The entire piece is called a rib.

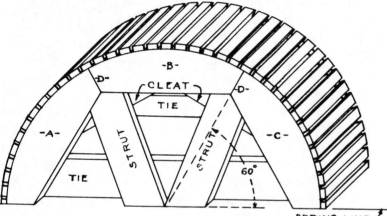
Parts Of An Arch Center
Figure 14-12

If the width of the jamb or wall is 4 inches or less, only one rib will be necessary. If the jamb is 4 inches to 17 inches, two ribs should be used. Figure 14-13 shows the finished arch with the center removed.

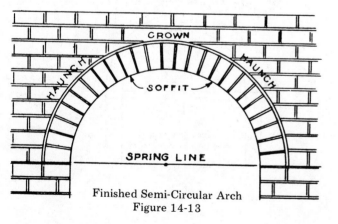
Finished Semi-Circular Arch
Figure 14-13

HOW TO BUILD THE SEMI-CIRCULAR ARCH CENTER

1. Lay out the curve of the arch with trammel points on the subfloor. Use a radius of one half the width of the opening minus the thickness of the lagging (Figure 14-14).

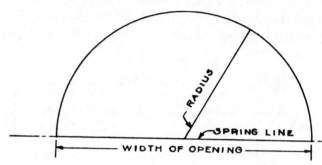
Layout Of Semi-Circular Arch
Figure 14-14

2. Cut three pieces of wood and tack them to the floor over the arch mark (Figure 14-16). The outside pieces (F and G) should be on top and the outside edge of each board should cover the curved line. The width of the boards where the angle cuts are to be made should be not less than 3 inches.

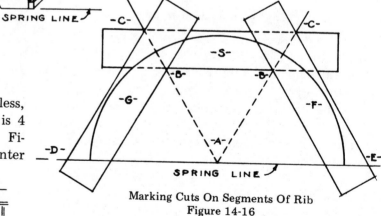
Marking Cuts On Segments Of Rib
Figure 14-16

3. Draw two straight lines from the center point A, Figure 14-16, through the intersecting edges (C and B) of the boards. These lines show the miter cuts on the boards F and G. The line DE marks the bottom angle cuts on these same boards.

4. Make the cuts on both ends of boards F and G.

112

5. Re-set the pieces F and G and mark the top piece S.

6. Remove this piece and make the cuts on the ends.

7. Re-set the three pieces as at Figure 14-17. Check to see that the three pieces are on the curved line and at the spring line.

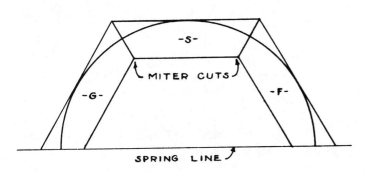

Layout Of Semi-Circle On Rib
Figure 14-17

8. Strike the arc on these pieces (Figure 14-17). Cut along the curved line of each piece with a compass saw and finish with a spoke shave or plane.

9. Re-set each piece in place and nail cleats over the joints (Figure 14-18). These cleats should be at least 16 inches long. Be sure they do not project above the top of the rib.

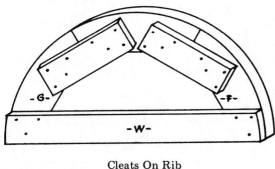

Cleats On Rib
Figure 14-18

10. Cut a tie for the bottom (W, Figure 14-18) and nail this to the two bottom segments (F and G). The ends of the tie should be slightly curved to match the curve of the segments.

NOTE: If the center is over 4 feet wide, it will be necessary to use struts. They should be inserted at the miter cuts where they will strengthen the weakest points of the rib.

11. Turn the rib over and cut in the struts X and Y, Figure 14-19. These can be nailed to the cleats and the tie. The other rib can be made in the same manner.

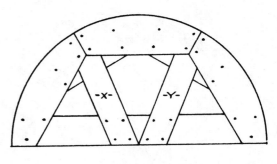

Struts On Rib
Figure 14-19

NOTE: Should it be necessary to build several arch centers of the same dimensions, make one rib, check it for accuracy, and then use this rib as a pattern for all ribs of all the arches. This same method should be followed in making the struts, lags, cleats, and braces.

12. Cut the required amount of lagging.

13. Tack the two ribs of the arch center to the floor (Figure 14-20). Nail a piece of lagging across each end of the pair of ribs to space them the proper distance apart.

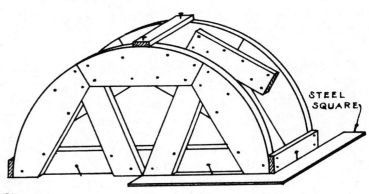

Attaching Lagging
Figure 14-20

14. Tack one piece at the top temporarily, so the two ribs will be spaced correctly.

15. With a steel square, check the two ribs for squareness (Figure 14-20) and make adjustments if necessary. Nail on the remainder of the lagging. Be sure that it is nailed square with the ribs. Figure 14-21 shows how the center should look when complete.

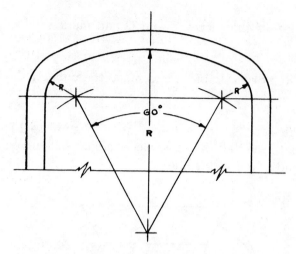

Layout Of Semi-Elliptical Arch
Figure 14-22

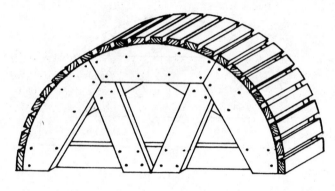

Completed Arch Center
Figure 14-21

The finished arch with the center removed is shown in Figure 14-24.

NOTE: The semi-elliptical arch center is made in the same way as the segmental arch center. Only the layout, as shown above, is different.

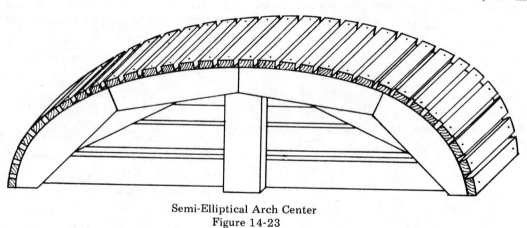

Semi-Elliptical Arch Center
Figure 14-23

The Semi-Elliptical Arch. This type of arch center is one half of an ellipse and is laid out as shown in Figure 14-22. It is made somewhat like the semi-circular arch center but it is more difficult to construct because it requires a number of pieces to form the curve (Figure 14-23). If the arch is wide and the masonry heavy, the cleats holding the segments together are made continuous on the back. Struts are used if the arch is over 3 feet wide. These struts should be placed at the joints since they are the weakest points of the arch center.

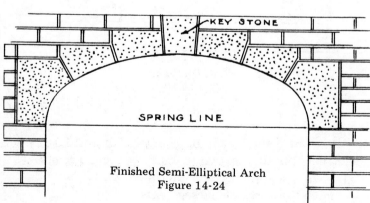

Finished Semi-Elliptical Arch
Figure 14-24

114

HOW TO SET ARCH CENTERS

NOTE: Figure 14-25 shows a doorway arch center set in place before the finished door frame is installed. Centers for window frames are sometimes set in conjuction with the window frames. In this case, the centers rest on the frames. If the frames have not been set, the centers will have to be supported by shores.

1. Cut four posts or shores long enough to support the center from the floor. See A and C, Figure 14-25.

2. Set one of these shores upright against each corner of the opening as shown by A, B, C, D, Figure 14-25.

3. Cut two pieces of 2 x 4 to the same length as the thickness of the masonry wall. See E and F, Figure 14-25.

4. Nail the 2 x 4 plates on top of the shores. These act as ties for the shores and also as nailing surfaces for the arch center.

5. Cut four pieces 1 inch x 6 inches for cleats to tie and brace the shores (G, Figure 14-25). Nail these in place.

6. Cut and place a pair of wedges under each shore. These are to be used to bring the top of the shores up to the spring line of the masonry arch, and to simplify the removal of the center after the

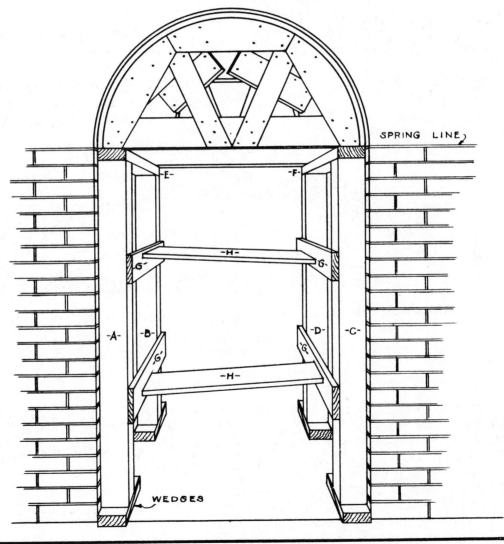

Arch Center In Place
Figure 14-25

115

mortar in the arch has set.

7. Cut two braces (H) to such a length that they will wedge the shores tightly against the sides of the opening. Toenail these in place.

8. Set the center in place and level it by adjusting the wedges at the bottom of the shores. Toenail the wedges and arch center in place.

NOTE: If there is any danger of the center tipping, nail a brace from the under side of the top lagging to the floor or shore post. Do not nail this brace to the scaffold as it would then move if workmen walked on the planks.

CHAPTER 15

HANDLING AND PLACING CONCRETE

Each step in handling and transporting of concrete must be carefully controlled to maintain uniformity within the batch and from batch to batch so that the completed work is consistent throughout. The method of handling and transporting concrete and the equipment used should not place restrictions on the consistency of the concrete. Consistency should be governed by the placing conditions. If these conditions allow a stiff mix, then equipment should be chosen which is capable of handling and transporting such a mix.

Requirements. The three main requirements for transportation of concrete from the mixing plant to the job sites are:

(1) It must be rapid so that the concrete will not dry out or lose its workability or plasticity between mixing and placing.

(2) Segregation of the aggregates and paste must be reduced to a minimum to assure uniform concrete. Loss of fine material, cement or water should be prevented.

(3) Transportation should be organized so that there are no undue delays in the placing of concrete for any particular unit or section that would cause undesirable fill planes or construction joints.

Handling Techniques. Several general points concerning the handling of concrete are illustrated in Figure 15-1. Failure to observe the right procedures illustrated to prevent the segregation of the aggregates and paste can result in poor concrete in spite of good design and mixing procedures. Separation or segregation occurs because concrete is made up of materials of different particle size and specific gravity. The coarser particles in a concrete mix placed in a bucket tend to settle to the bottom and the water rises to the top. Honeycomb concrete or rock pockets are caused by the segregation of materials.

DELIVERY METHODS

Equipment provided for the delivery of concrete should be based on 100-percent of the anticipated placing rate. When transit mix equipment will not have access to the site, wheelbarrows or buggies are usually the most practical and economical means of delivering concrete for foundations, foundation walls, or slabs poured on or below grade. Power buggies, if available, are useful for longer runs. Hand buckets may be used to deliver small quantities of concrete as the occasion demands. When a situation arises requiring the concrete to be poured approximately 15 feet above grade, inclined runway (2, Figure 15-2) can be constructed economically so that buggies or wheelbarrows can be used. When concrete is deposited below or at approximate grade, 2-inch plank runways placed on the ground permit the concrete to be poured directly into the form. If the difference in elevation from the runway to the bottom of the structure to be poured is large a chute similar to the illustration (Figure 15-3) should be used to avoid segregation. Usually the slope of the chute should be 2:1 or steeper for stiff mixes. If concrete is to be transported by buggy or wheelbarrow, suitable runway must be provided. A runway along a wall form is shown in 1, Figure 15-2. The top of the runway should be about level with the top of the form sheathing. In order to provide room for the ledger, the top wale should be kept at least 1 foot below the top of the concrete. The runway should be made from rough lumber and should consist of 2- by 10-inch planks supported by 4 by 4's spaced on 6-foot centers. The 1- by 6-inch ledger on the form side should be nailed to the studs on the wall form. Runways must always be securely braced in order to prevent failure. Concrete can be elevated to about 5 to 6 feet above the level of the mixer by wheeling the concrete up inclined runways that have a slope of 10:1. This type of runway is shown in 2, Figure 15-2. If the concrete must be lifted more than 5 to 6 feet, and a large quantity is involved, it is more economical to use elevating equipment, such as a bucket and a crane. Runways for placing concrete on a floor slab should be supported from the form work or from the ground. These runways should also consist of 2 by 10's supported by 4 by 4's placed on 6-foot centers. If possible, all runways should be arranged so that the buggies or wheelbarrows will not have to pass each other on the runways at any time.

117

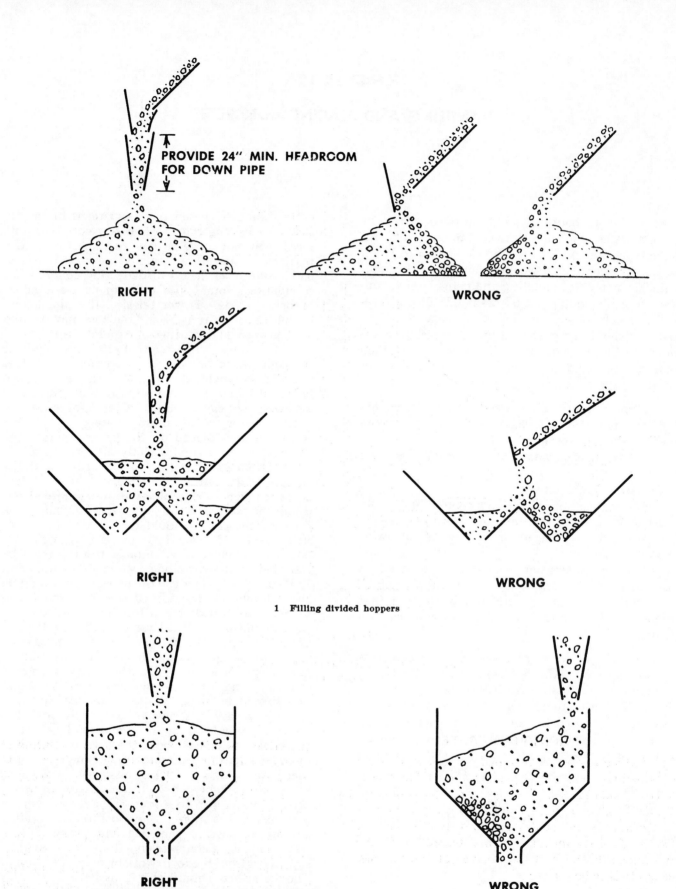

PROVIDE 24" MIN. HEADROOM
FOR DOWN PIPE

RIGHT

WRONG

RIGHT

WRONG

1   Filling divided hoppers

RIGHT

WRONG

2   Filling hoppers or buckets

Concrete Handling Techniques
Figure 15-1

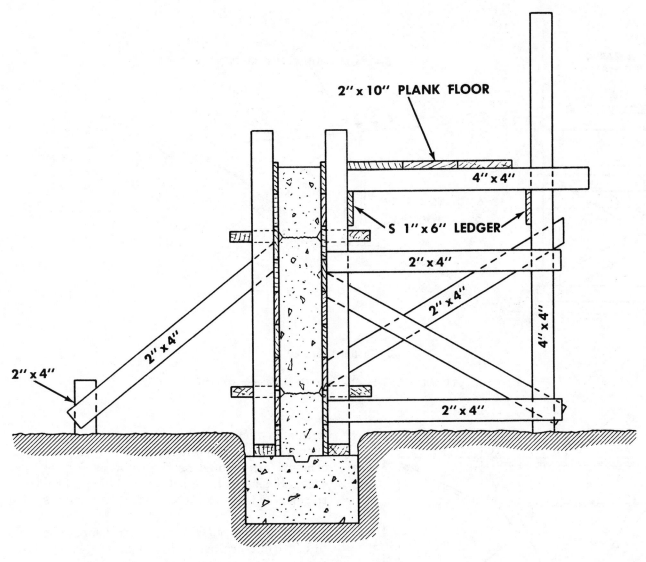

2" x 10" PLANK FLOOR

4" x 4"

S 1" x 6" LEDGER

2" x 4"

2" x 4"

4" x 4"

2" x 4"

2" x 4"

2" x 4"

1   Runway along a wall

Runways For Use Of Wheelbarrows Or Buggies
Figure 15-2

PLACING CONCRETE

The full value of well-designed concrete cannot be obtained without proper placing and curing procedures. Good concrete placing and compacting techniques produce a tight bond between mortar and coarse aggregate and assure complete filling of the forms. These requirements are necessary if the full strength and best appearance of the finished concrete is to be realized.

PRELIMINARY PREPARATION

General Preparation. Preparation prior to concreting includes compacting, trimming, and moistening the subgrade; erecting the forms; and setting the reinforcing steel. A moist subgrade is especially important in hot weather to prevent extraction of water from the concrete.

Rock Subgrades. When rock must be cut out, the surfaces in general should be vertical or horizontal rather than sloping. The rock surface should be roughened and thoroughly cleaned. Stiff brooms, water jets, high-pressure air, or wet sandblasting may be used. All water depressions should be removed and the rock surface coated

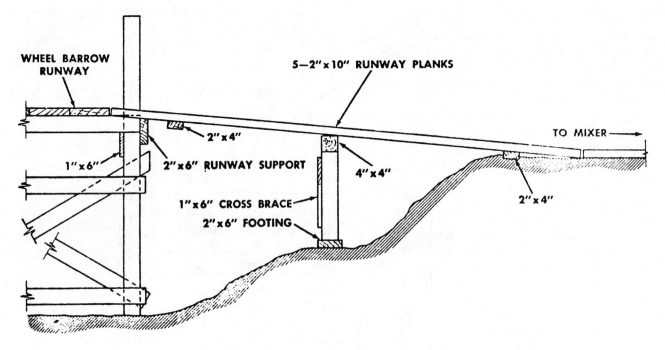

2   Inclined runway

Runways For Use Of Wheelbarrows Or Buggies — continued
Figure 15-2

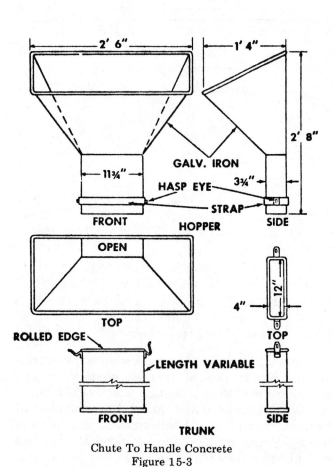

Chute To Handle Concrete
Figure 15-3

with a ¾-inch thick layer of mortar. The mortar should contain only fine aggregate and the water-cement ratio should be the same as for the concrete. It should have a 6-inch slump. Then work the mortar into the surface with stiff brushes.

Clay Subgrades. A subgrade composed of clay or other fine-grained soils should be moistened to a depth of 6 inches to aid in curing the concrete. If the soil is sprinkled intermittently, it can be saturated without becoming muddy. The surface must be clean of debris and any dry loose material before concrete is placed.

Gravel and Sand Subgrades. Subgrades composed of gravel or other loose material need a tar paper or burlap cover before concrete is placed on them. Compacted sand need not be covered. However, it must be moist when the concrete is placed to prevent the absorption of water from the concrete. Tar paper should be lapped not less than 1 inch, and stapled. Burlap should be joined by sewing with wire and moistened by sprinkling before concrete is placed on it.

Preparation of Forms. Shortly before concrete is placed, the forms should be checked for tightness and cleanliness. Bracing should be checked to assure that there will be no movement of the forms during placing. The forms are coated with a suitable form oil or coating material that will keep the concrete from sticking. In an emergency, the forms can be moistened with water to prevent sticking. Forms that have been exposed to the sun for some time will dry out and the joints will tend to open up. Saturating the forms with water helps to close such joints.

Depositing Fresh Concrete on Hardened Concrete. When new concrete is deposited on hardened concrete, the hardened concrete must be nearly level, clean, and moist, with some aggregate particles partially exposed, to obtain a good bond and a watertight joint. If there is a soft layer of mortar or laitance on the surface of the hardened concrete it should be removed. Wet sand blasting and washing is the most effective means of preparing the old surface where the sand deposit can be easily removed. Always moisten hardened concrete before placing new concrete. Dried-out concrete must be saturated for several hours. In no case should there be pools of water on the old surface when fresh concrete is deposited.

DEPOSITING CONCRETE

Basic Considerations. Concrete should be deposited in even horizontal layers and should not be puddled or vibrated into place. The layers should be from 6 to 24 inches in depth depending on the type of construction. The initial set should not take place before the next layer is added. To prevent honeycombing or avoid spaces in the concrete, the concrete should be vibrated or spaded. This is particularly desirable in wall forms with considerable reinforcing. Care should be taken not to overvibrate, because segregation and a weak surface may result.

(1) *Maximum drop.* There is a temptation, in the interests of time and effort, to drop the concrete from the point to which it has been transported regardless of the height of the forms, but the free fall of concrete into the forms should be reduced to a maximum of 3 to 5 feet unless vertical pipes, suitable drop chutes, or baffles are provided. Figure 15-1 suggests several methods to control the fall of concrete and prevent honeycombing of concrete and other undesirable results.

(2) *Thickness of layers.* Concrete should be deposited in horizontal layers whenever possible and each layer consolidated before the succeeding layer is placed. Each layer should be placed in one operation. In mass concrete work, where concrete is deposited from buckets, the layers should be from 15 to 20 inches thick. For reinforced concrete members the layers should be from 6 to 20 inches thick. The thickness of the layers depends on the width between forms and the amount of reinforcement.

(3) *Positioning.* Concrete should be placed as nearly as possible in its final position. Horizontal movement should be avoided since this results in segregation because mortar tends to flow ahead of coarser material. Concrete should be worked thoroughly around the reinforcement and bedded fixtures, into the corners, and on the sides of the forms.

(4) *Rate of placing.* On large pours so as to avoid excess pressure on forms, the rate of filling should not exceed 4 feet per hour measured vertically, except for columns. Placing will be coordinated so that the concrete is not deposited faster than it can be properly compacted. In order to avoid cracking during settlement an interval of at least 4 hours, preferably 24 hours, should elapse between completion of columns and walls and the placing of slabs, beams, or girders supported by them.

Wall Construction. For walls, the first batches should be placed at the ends of the section. Placing should then proceed toward the center for each layer, if more than one layer is necessary, to prevent water from collecting at the ends and corners of the forms. This method should also be used in placing concrete for beams and girders. For wall construction, the inside form should be stopped off at the level of the construction. Overfill the form for about 2 inches and remove the excess just before setting occurs to insure a rough, clean surface. Before the next lift of concrete is placed on this surface, a ½ to 1-inch thick layer of sand-cement mortar should be deposited on it. The mortar should have the same water content as the concrete and should have a slump of about 6 inches to prevent stone pockets and help produce a water tight joint. Proper procedure for placing concrete in walls is shown in 1, Figure 15-4. Note the use of drop chutes and port openings for placing concrete in the lower portion of the wall. The port openings are located at about 10-foot intervals along the wall. Concrete for the top portion of the wall can be placed as shown in 2, Figure 15-4. When pouring walls, remove the form spreaders as the forms are filled.

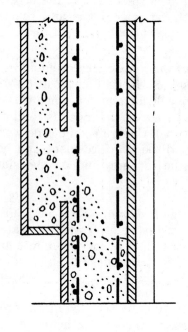

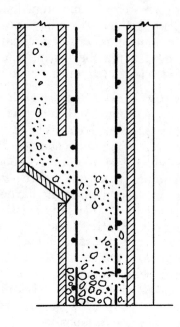

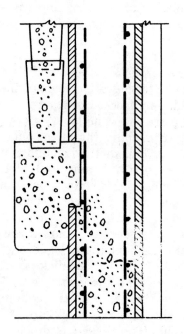

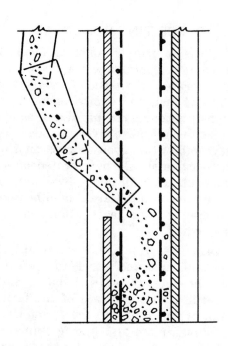

RIGHT   1   Placing concrete in high wall form   **WRONG**

Concrete Placing Techniques
Figure 15-4

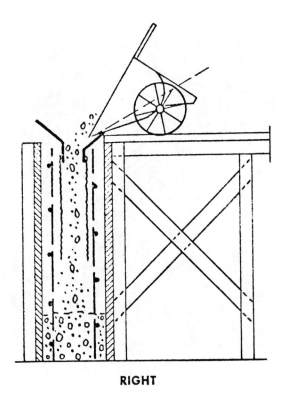
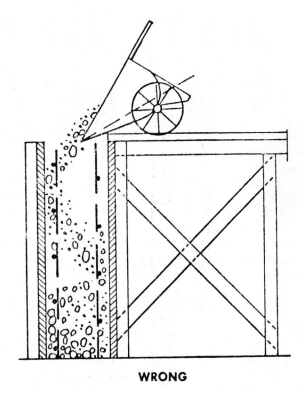

RIGHT **WRONG**

2   Placing concrete in top of form

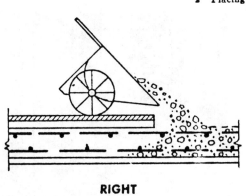
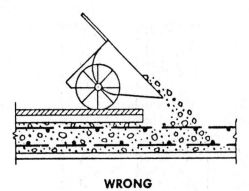

RIGHT **WRONG**

3   Placing concrete in slab

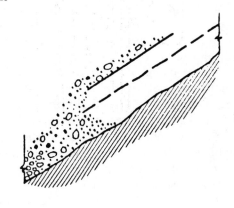

RIGHT **WRONG**

4   Placing concrete in a sloping surface

Concrete Placing Techniques - continued
Figure 15-4

Slab Construction. For slabs, the concrete should be placed at the far end of the slab, each batch dumped against previously placed concrete, as shown in 3, figure 15-4. The concrete should not be dumped in separate piles and the piles then leveled and worked together. Nor should the concrete be deposited in big piles and then moved horizontally to its final position, since this practice results in segregation.

Placing Concrete on Slopes. Procedure for placing concrete on slopes is shown in 4, figure 15-4. Always deposit the concrete at the bottom of the slope first, and proceed up the slope as each batch is dumped against the previous one. Compaction is thus increased by the weight of the newly added concrete when it is consolidated.

CONSOLIDATING CONCRETE

Purpose. With the exception of concrete placed under water, concrete is compacted or consolidated after placing. Consolidation may be accomplished by the use of hand tools such as spades, puddling sticks, and tampers; but the use of mechanical vibrators is preferred. Compacting devices must reach the bottom of the form and must be small enough to pass between reinforcing bars. Consolidation eliminates rock pockets and air bubbles and brings enough fine material to the surface and against forms to produce the desired finish. In the process of consolidation the concrete is carefully worked around all reinforcing steel to assure proper embedding of the steel in the concrete. Displacement of reinforcing steel must be avoided since the strength of the concrete member depends on proper location of the reinforcement.

Vibration. Consolidation is effectively accomplished by use of mechanical vibrators, as shown in figure 15-5. Vibrators consolidate concrete by pushing the coarse aggregate down and away from the point of vibration. With vibrators it is possible to place concrete mixtures too stiff to be placed in any other way. In most structures, concrete with a 1- or 2-inch slump can be deposited. Stiff mixtures require less cement and are therefore more economical. Moreover, there is less danger of segregation and excessive bleeding. The mix must not be so stiff that an excessive amount of labor is required to place it.

The internal vibrator involves insertion of a vibrating element into the concrete. The external type is applied to the forms. It is powered by electric motor, gasoline engine, or compressed air. The internal vibrator should be inserted in the con-

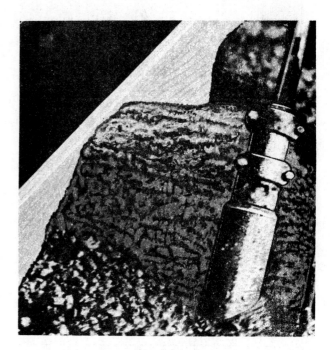

1 Before vibration

2 After vibration

Use Of A Vibrator To Consolidate Concrete
Figure 15-5

crete at intervals of approximately 18 inches for 5 to 15 seconds to allow some overlap of the area vibrated at each insertion. Whenever possible the vibrator should be lowered vertically into the concrete and allowed to descend by gravity. The vibrator should pass through the layer being placed and penetrate the layer below for several inches to insure a good bond between the layers. Under normal conditions there is little likelihood of damage from the vibration of lower layers provided the disturbed concrete in these lower layers becomes plastic under the vibratory action. Sufficient vibration has taken place when a thin line of mortar appears along the form near the vibrator, when the coarse aggregate has sunk into the concrete, or when the paste just appears near the vibrator head. The vibrator should then be withdrawn vertically at about the same rate that it descended. The length of time that a vibrator should be left in the concrete depends on the slump of the concrete. Mixes that can be easily consolidated by spading should not be vibrated because segregation may occur. Concrete that has a slump of 5 or 6 inches should not be vibrated. Vibrators should not be used to move concrete any distance in the form. Some hand spading or puddling should accompany vibration.

Hand Methods. Hand methods for consolidating concrete include the use of spades or puddling sticks and various types of tampers. For consolidation by spading, the spade should be shoved down along the inside surface of the forms through the layer deposited and down into the lower layer for a distance of several inches, as shown in figure 15-6. Spading or puddling should continue until the coarse aggregate has disappeared into the mortar.

PLACING CONCRETE UNDER WATER

Suitable Conditions. Concrete should be placed in air rather than under water whenever possible. When it must be placed under water, the work should be done under experienced supervision and certain precautions should be taken. For best results, concrete should not be placed in water having a temperature below 45° F. and should not be placed in water flowing with a velocity greater than 10 feet per minute, although sacked concrete may be used for water velocities greater than this. If the water temperature is below 45° F., the temperature of the concrete when it is deposited should be above 60° F. but in no case above 80° F. If the water temperature is above 45° F., no tem-

perature precautions need be taken. Coffer dams or forms must be tight enough to reduce the current to less than 10 feet per minute through the space to be concreted. Pumping of water should not be permitted while concrete is being placed or for 24 hours thereafter.

Tremie Method. Concrete can be placed under water by several methods, the most common of which is with a tremie. The tremie method involves a device shown in figure 15-7. A tremie is a pipe having a funnel-shaped upper end into which the concrete is fed. The pipe must be long enough to reach from a working platform above water level to the lowest point at which the concrete is to be deposited. Frequently the lower end of the pipe is equipped with a gate, permitting filling before insertion in water. This gate can be opened from above at the proper time. The bottom or discharge end is kept continuously buried in newly placed

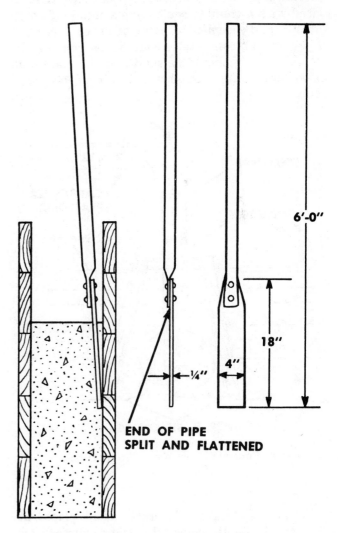

Consolidation By Spading And The Spading Tool
Figure 15-6

125

concrete, and air and water are excluded from the pipe by keeping it constantly filled with concrete. The tremie should be lifted slowly to permit the concrete to flow out. Care must be taken not to lose the seal at the bottom. If lost, it is necessary to raise the tremie, plug the lower end, and lower the tremie into position again. The tremie should not be moved laterally through the deposited concrete. When it is necessary to move the tremie, it should be lifted out of the concrete and moved to the new position, keeping the top surface of the concrete as level as possible. A number of tremies should be used if the concrete is to be deposited over a large area. They should be spaced on 20- to 25-foot centers. Concrete should be supplied at a uniform rate to all tremies with no interruptions at any of them. Pumping from the mixer is the best method of supplying the concrete. Large tremies can be suspended from a crane boom and can be easily raised and lowered with the boom. Concrete that is placed with a tremie should have a slump of about 6 inches and a cement content of seven sacks per cubic yard of concrete. About 50 percent of the total aggregate should be sand and the maximum coarse aggregate size should be from 1½ to 2 inches.

Concrete placed by bucket can be slightly stiffer than that placed by tremie but it should still contain seven sacks of cement per cubic yard. The bucket is completely filled and the top covered with a canvas flap. The flap is attached to one side of the bucket only. The bucket is lowered slowly into the water so that the canvas will not be displaced. Concrete must not be discharged from the bucket before the surface upon which the concrete is to be placed has been reached. Soundings should be made frequently so that the top surface is kept level.

Sacked Concrete. In an emergency, concrete can be placed under water in sacks. Jute sacks of about 1-cubic foot capacity, filled about two-thirds full, are lowered into the water, preferably shallow water. These sacks are placed in header and stretcher courses, interlocking the entire mass. A header course is placed so that the length of the sack is at right angles to the direction in which the stretcher-course sacks are laid. Cement from one sack seeps into adjacent sacks and they are thus bonded together. Experience has shown that the less the concrete under water is disturbed after placement, the better it will be. For this reason, compaction should not be attempted.

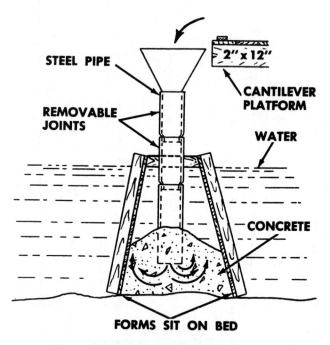

Placing Concrete Under Water With A Tremie
Figure 15-7

Concrete Buckets. Concrete can be placed at considerable depth below the water surface by means of open-top buckets with a drop bottom.

CHAPTER 16

FINISHING CONCRETE

The finishing process provides the desired surface effect of the concrete. The concrete finishing process may be performed in many ways, depending on the effect required. Occasionally only correction of surface defects, filling of bolt holes or cleaning is necessary. Unformed surfaces may require only screeding to proper contour and elevation, or a broomed, floated, or troweled finish may be specified.

Screeding. After a floor slab, sidewalk or pavement has been placed, the top surface is rarely at the exact elevation desired. The process of striking off the excess concrete in order to bring the surface to the right elevation is called screeding. This operation can begin as soon as the concrete has been placed. Prior to screeding the concrete should be vibrated to lower larger sized aggregate to avoid interference with the screed. A templet with a straight lower edge if a flat surface is required, or curved if a curved surface is required, is moved back and forth across the concrete with a sawing motion. The templet rides on wood or metal strips that have been established as guides. With each sawing motion the templet is moved a short distance along the forms as shown in figure 16-1. There should be a surplus of concrete against the front face of the templet which will be forced into the low spots as the templet is moved forward. If there is a tendency for the templet to tear the surface, the rate of forward movement of the templet should be reduced or the bottom edge should be covered with metal. In most cases this will stop the tearing action. Such procedures are necessary when air-entrained concrete is used because of the sticky nature of this type of concrete. It is possible to hand screed surfaces up to 30 feet in width but for efficient screeding it is best not to go beyond 10 feet. Three men, excluding a vibrator operator, can screed approximately 200 square feet of concrete per hour. Two men operate the screed and the third man pulls excess concrete from the front of the screed. It is necessary to screed the surface twice to remove the surge of excess concrete caused by the first screeding.

Screeding Operations
Figure 16-1

Floating. If a smoother surface is required than the one obtained by screeding, the surface should be worked sparingly with a wood or metal float or finishing machine. A wood float is shown in 1, figure 16-2 and the float in use is shown in 2, figure 16-2. This process should take place shortly after screeding and while the concrete is still plastic and workable. Floating should not begin until the water sheen has disappeared and the concrete has hardened sufficiently that a man's foot leaves only a slight imprint. The purpose of floating is threefold: to embed aggregate particles just beneath the surface; to remove slight imperfections, high spots, and low spots; and to compact the concrete at the surface in preparation for other finishing operations. The concrete must not be overworked while it is still plastic, to avoid bringing an excess of water and mortar to the surface. This fine material will form a thin weak layer that will scale or wear off under usage. Where a coarse texture is desired as the final finish, it is usually necessary to float the surface a second time after it has partially hardened so that the required surface will be obtained. In slab construction long-handled wood floats are used as shown in 3, figure 16-2. The steel float is used the same way as the wood float but it gives the finished concrete a much smoother surface. Steel floating should begin when the water sheen disappears from the concrete surface, to avoid cracking and dusting of the finished concrete. Cement or water should not be used to aid in finishing the surface.

1   Wood float

2   Floating operation

3   Long-handling wood float and floating operation

Wood Floats And Floating Operations
Figure 16-2

Troweling. If a dense, smoother finish is desired, floating must be followed by steel troweling at some time after the moisture film or sheen disappears from the floated surface and when the concrete has hardened enough to prevent fine material and water from being worked to the surface. This step should be delayed as long as possible. Excessive troweling too early tends to produce crazing and lack of durability; too long a delay in troweling results in a surface too hard to finish properly. The usual tendency is to start to trowel too soon. Troweling should leave the surface smooth, even, and free of marks and ripples. Spreading dry cement on a wet surface to take up excess water is not good practice where a wear-resistant and durable surface is required. Wet spots must be avoided if possible; when they do occur, finishing operations should not be resumed until the water has been absorbed, has evaporated, or has been mopped up. A surface that is fine-textured but not slippery may be obtained by troweling lightly over the surface with a circular motion immediately after the first regular troweling. In this process, the trowel is kept flat on the surface of the concrete. Where a "hard steel-troweled finish" is required, the first regular troweling is followed by a second troweling after the concrete has become hard enough so that no mortar ad-

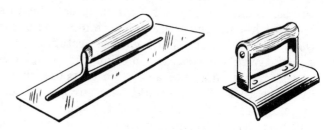

1   Steel trowel and edger

2   Troweling operation

Steel Finishing Tools And Troweling Operations
Figure 16-3

heres to the trowel and a ringing sound is produced as the trowel passes over the surface. During this final troweling, the trowel should be tilted slightly and heavy pressure exerted to thoroughly compact the surface. Hair cracks are usually due to a concentration of water and fines at the surface resulting from overworking the concrete during finishing operations. Such cracking is aggravated by too rapid drying or cooling. Cracks that develop before troweling usually can be closed by pounding the concrete with a hand float. A steel trowel and an edger are shown in 1, figure 16-3 and the troweling operation in 2, figure 16-3.

Brooming. A nonskid surface can be produced by brooming the concrete before it has thoroughly hardened. Brooming is carried out after the floating operation. For some floors and sidewalks where severe scoring is not desirable, the broomed finish can be produced with a hair brush after the surface has been troweled to a smooth finish once. Where rough scoring is required, a stiff broom made of steel wire or coarse fiber should be used. Brooming should be done in such a way that the direction of the scoring is at right angles to the direction of the traffic.

Rubbed Finish. A rubbed finish is required when a uniform and attractive surface must be obtained although it is possible to produce a surface of satisfactory appearance without rubbing if plywood or lined forms are used. The first rubbing should be done with coarse carborundum stones as soon as the concrete has hardened so that the aggregate is not pulled out. The concrete should then be cured until final rubbing. Finer carborundum stones are used for the final rubbing. The concrete should be kept damp while being rubbed. Any mortar used to aid in this process and left on the surface should be kept damp for 1 to 2 days after it sets in order to cure properly. The mortar layer should be kept to the minimum as it is likely to scale off and mar the appearance of the surface.

Machine Finishing. Machine finishing is carried out at such time as the concrete takes its initial set. The concrete must, however, be in workable condition at the time of the finishing operation. The screeds and vibrator on the machine finisher are set to give the proper surface elevation and produce a dense concrete. In most cases, there should be a sufficiently thick layer of mortar ahead of the screed to insure that all low spots will be filled. The vibrator follows the front screed and the rear screed is last. The rear screed should be adjusted to carry enough grout ahead of it to insure continuous contact between screed and pavement. If forms have been set in good alignment and firmly supported, and if the concrete has the right workability, no more than two passes of the machine should be required to produce a satisfactory surface.

REPAIRING CONCRETE

After forms are removed, small projections must be removed, tierod holes filled and honeycombed areas repaired. These repairs should be made as soon as possible after the forms are removed. The repair of concrete is covered in detail in Chapter 17.

CLEANING CONCRETE

Concrete surfaces are not always uniform in color when forms are removed. If appearance is important the surface should be cleaned.

The surface can be cleaned with a cement-sand mortar consisting of one part portland cement and one and one-half to two parts fine sand. The mortar should be applied to the surface by a brush after all defects have been repaired. If a light-colored surface is desired, white portland cement can be used. The surface should be scoured vigorously with a wood or cork float immediately after applying the mortar. Excess mortar should be removed with a trowel after 1 or 2 hours which gives time for the mortar to harden enough so the trowel will not remove it from the small holes. After the surface has dried, it should be rubbed with dry burlap to remove any loose material. No visible film of mortar should remain after the rubbing. One section should be completed without stopping. Mortar left on the surface overnight is very difficult to remove.

An alternate method of cleaning with mortar consists of rubbing the mortar over the surface with clean burlap. The mortar should have the consistency of thick cream and the surface should be almost dry. The excess mortar is then removed by rubbing with clean burlap. Removal should be delayed long enough to prevent smearing but should be completed before the mortar hardens. The mortar is allowed to set several hours, then cured for 2 days. After curing, the surface is permitted to dry and is vigorously sanded with No. 2 sandpaper. This removes all excess mortar not removed by the sack rubbing and produces a surface of uniform appearance. For best results, mortar cleaning should be done in the shade on a cool, damp day.

Sandblasting. Surface stains, particularly rust, can be completely removed by lightly sandblasting the surface. This method is more effective than washing with acid.

Acid Cleaning. Acid washing can be used where the staining is not severe. Acid washing should be preceded by a two-week period of moist curing. The surface is first wetted and, while still damp, is scrubbed thoroughly with a 5 to 10 percent solution of muriatic acid using a stiff bristle brush. The acid is removed by immediate, thorough flushing with clean water. If possible, acid washing should be followed by four additional days of moist curing. When handling the acid, goggles are worn to protect the eyes and precautions must be taken to prevent the acid from contacting hands, arms, and clothing.

CHAPTER 17

CURING AND PATCHING CONCRETE

Hydration. The addition of water to portland cement and the formation of a water-cement paste starts a chemical reaction whereby the cement becomes a bonding agent. During this process, known as hydration, the main compounds of portland cement and water react to form products of hydration which produce a firm and hard substance—the hardened cement paste. The rate and degree of hydration, and as a result the strength of the concrete, are dependent on the curing process followed after the concrete has been placed and consolidated. The process of hydration continues for an indefinite period at a decreasing rate as long as water is in the mixture and temperature conditions are favorable. Once the water is removed, hydration ceases and cannot be restarted.

Importance. Curing refers to the steps necessary to keep concrete moist and as near to 73° F. as practicable until it has reached its design strength. Properties of concrete such as resistance to freezing and thawing, strength, watertightness, wear resistance, and volume stability improve with age as long as these conditions, which are favorable for continued hydration, are maintained. It follows that concrete should be protected so that moisture is not lost during the early hardening period and that the concrete temperature is kept favorable for hydration.

Length of curing period. The length of

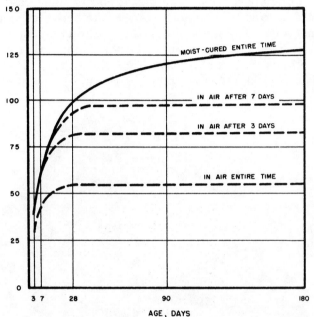

Effect Of Moist Curing On Concrete
Compressive Strength
Figure 17-1

Method	Advantage	Disadvantage
Sprinkling with water or covering with wet burlap.	Excellent results if constantly kept wet.	Likelihood of drying between sprinklings. Difficult on vertical walls.
Straw	Insulator in winter	Can dry out, blow away, or burn.
Moist earth	Cheap, but messy	Stains concrete. Can dry out. Removal problem.
Ponding on flat surfaces	Excellent results, maintains uniform temperature.	Requires considerable labor, undesirable in freezing weather.
Curing compounds	Easy to apply. Inexpensive	Sprayer needed. Inadequate coverage allows drying out. Film can be broken or tracked off before curing is completed. Unless pigmented, can allow concrete to get too hot.
Waterproof paper	Excellent protection, prevents drying	Heavy cost can be excessive. Must be kept in rolls, storage and handling problem.
Plastic film	Absolutely watertight, excellent protection. Light and easy to handle.	Should be pigmented for heat protection. Requires reasonable care and tears must be patched. Must be weighed down to prevent blowing away.

Curing Methods
Table 17-1

time that concrete should be protected against loss of moisture is dependent upon the type of cement, mix proportions, required strength, size and shape of the concrete mass, weather, and future exposure conditions. The period may vary from a few days to a month or longer. The influence of curing on the strength of concrete is shown in figure 17-1.

CURING METHODS

Concrete can be kept moist, and in some cases at a favorable temperature by a number of curing methods. These may be classified into two categories: methods that supply additional moisture, and methods that prevent loss of moisture. Table 17-1 lists several effective methods of curing concrete

Curing A Wall With Wet Burlap Sacks
Figure 17-2

Use Of Waterproof Curing Paper
Figure 17-3

together with comparable advantages and disadvantages.

Methods that Supply Additional Moisture. These methods add moisture to the surface of the concrete during the early hardening period. Such methods include sprinkling and wet coverings. Some cooling, through evaporation, is provided which is beneficial in hot weather.

Sprinkling. Continuous sprinkling with water is an excellent method of curing. If sprinkling is done at intervals, the concrete must not be allowed to dry out between applications. The expense involved and volume of water required may be disadvantageous.

Wet coverings. Wet coverings such as burlap, cotton mats, and other moisture retaining fabrics are used extensively for curing. Straw and moist earth may also be used. These coverings should be placed as soon as the concrete has hardened sufficiently to prevent surface damage. A typical use of wet burlap is shown in figure 17-2. Wet coverings should remain in place and be kept moist during the entire curing period. For most structural use, the curing period for cast in place concrete is usually 3 days to 2 weeks, depending upon such conditions as temperature, cement type, mix proportions, etc. More extended curing periods are desirable for bridge decks and other slabs exposed to weather and chemical attack.

Methods that Prevent Loss of Moisture. These methods prevent loss of moisture by sealing the surface. This may be done by means of waterproof paper, plastic film, liquid-membrane-forming compounds, and forms left in place.

(1) *Waterproof paper.* Waterproof paper is an efficient means of curing horizontal surfaces and structural concrete of relatively simple shapes. The paper should be large enough to cover the width and edges of the slab. Adjacent sheets should be lapped 12 inches or more and the edges should be weighted down, to form a continuous cover with completely closed joints. The surface should be wet with a fine spray of water before covering. The coverings should remain in place during the entire curing period. The use of waterproof curing paper is depicted in figure 17-3.

(2) *Plastic film.* Certain plastic sheet materials are used in curing concrete. They are lightweight, effective moisture barriers and are easily applied to either simple or complex shapes. In some cases, thin plastic sheets may discolor hardened concrete, especially if the surface has been steel-troweled to a hard finish. Coverage, overlap, weighting down of edges, and surface wetting requirements are similar to those for waterproof paper.

(3) *Curing compounds.* Curing compounds retard or prevent evaporation of moisture from the concrete. They are suitable not only for curing fresh concrete, but also for further curing of concrete after removal of forms or after initial moist curing. They are applied by spray equipment. Curing compound can be satisfactorily applied with hand operated pressure sprayers on odd widths or shapes of slabs and on concrete surfaces exposed by the removal of forms. Concrete surfaces subjected to heavy rain within 3 hours after curing compound is applied should be resprayed. Brushes may be used to apply the compound on formed surfaces, but should not be used on unformed concrete due to the danger of marring the concrete, opening the surface to excessive penetration of the compound, and breaking the continuity of the film. Curing compounds permit curing to continue for long periods while the concrete is in use. Curing compounds may prevent bond between hardened and fresh concrete, consequently they should not be used if bond is necessary.

(4) *Forms left in place.* Forms can provide adequate protection against loss of moisture if the top exposed concrete surfaces are kept wet. Wood forms left in place should be kept moist by sprinkling, especially during hot, dry weather.

REMOVING FORMS

It is generally advantageous to leave forms in place throughout the required curing period. However, it may be necessary to strip forms as early as possible to permit their immediate reuse. Also certain finishing operations, such as rubbing, may require early removal of forms. In any case, forms must not be removed before the concrete is strong enough to carry its own weight and any other loads that may be placed on it during construction. The forms for columns, footings, and sides of beams and walls can usually be removed before the forms for floors and beam bottoms. For most conditions, it is better to rely on the strength of the concrete as determined by test rather than to select arbitrarily the age at which forms may be removed. A minimum compressive strength of 500 psi should be attained before concrete is exposed to freezing. The age-strength relationship should be determined from tests on representative samples. Under average conditions (for example, air-entrained concrete made with a water-cement ratio of 6 gallons per sack) the times required to attain certain strengths are shown in table 17-2. It should be remembered that strengths are affected by materials used, temperatures, and other conditions. The time required before form removal, therefore, will vary from job to job.

Strength, psi	Age	
	Type I or normal	Type III or high-early-strength
500	24 hours	12 hours
750	1½ days	18 hours
1,500	3½ days	1½ days
2,000	5½ days	2½ days

*Air-entrained concrete with water-cement ratio of 6 gallons per sack.
Age-Strength Relationship*
Table 17-2

Forms should be designed and constructed with some thought as to their removal with a minimum of danger to the concrete. The forms must be stripped carefully to avoid damage to the surface of the concrete. When it is necessary to wedge against the concrete, only wood wedges should be used rather than a pinchbar or other metal tool. The forms should not be jerked off after wedging has been started at one end; this is almost certain to break the edges of the concrete. Forms that are to be reused should be cleaned and oiled immediately after their removal. Nails should be withdrawn as the forms are stripped from the concrete.

PATCHING

Inspection. Concrete should be inspected for surface defects when the forms are removed. These defects may be rock pockets, inferior quality, ridges at form joints, bulges, bolt holes, and form-stripping damage. Repairs are costly and interfere with the use of the structure. However, experience has demonstrated that no step in the procedure can be omitted or carelessly performed without harming the service ability of the repair work. If not properly performed, the repair will later become loose, will crack at the edges, and will not be water-tight.

Timely repair. On new work the repairs which will develop the best bond and thus have the best chance of being as durable and permanent as the original work are those made immediately after early stripping of the forms, while the concrete is quite green. For this reason, repairs should be performed within 24 hours after the forms have been removed.

Removal of ridges and bulges. If ridges and bulges are objectionable, they may be removed by careful chipping followed by rubbing with a grinding stone.

Defective areas. Defective areas such as rock pockets or honeycomb must be chipped out to solid concrete, the edges cut as straight as possible at right angles to the surface or slightly undercut to provide a key at the edge of the patch. The surface of all holes that are to be patched should be kept moist for several hours before applying the mortar. The mortar should be allowed to set as long as possible before being used, to reduce the amount of shrinkage and make a better patch. If a shallow layer of mortar is placed on top of the honeycombed concrete, moisture will form in the voids and subsequent weathering will cause the mortar to spall off. Shallow patches may be filled with mortar placed in layers not more than ½ inch thick. Each layer should be scratched rough to improve the bond with the succeeding layer, and the last layer smoothed to match the adjacent surface. Where absorptive form lining has been used, the patch can be made to match the rest of the surface by pressing a piece of the form lining against the fresh patch.

Large patches. Large or deep patches may be filled with concrete held in place by forms. These patches should be reinforced and doweled to the hardened concrete (figure 17-4). Patches usually

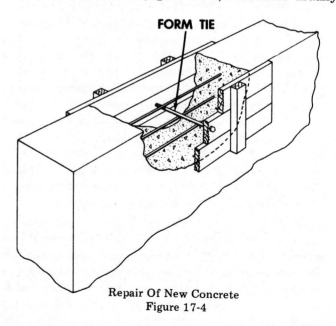

Repair Of New Concrete
Figure 17-4

appear darker than the surrounding concrete. Some white cement should be used in the mortar or concrete used for patching if appearance is important. A trial mix should be tried to determine the best proportion of white and gray cements to use. Before mortar or concrete is placed in patches, the surrounding concrete should be kept wet for several hours. A grout of cement and water mixed to a creamy consistency should then be brushed into the surfaces to which the new material is to be bonded. Curing should be started as soon as possible to avoid early drying. Damp burlap, tarpaulins and membrane curing compounds are useful for this purpose.

Filling bolt holes. Bolt holes should be filled with mortar carefully packed into place in small amounts. The mortar should be mixed as dry as possible, with just enough water so that it will be tightly compacted when forced into place. Tie rod holes extending through the concrete can be filled with mortar with a pressure gun similar to an automatic grease gun.

Flat surfaces. Feathered edges around a patch (1 figure 17-5) will break down. The chipped area should be at least 1 inch deep with the edges at right angles to the surface (2 figure 17-5). The correct method of screeding a patch is shown in 3 figure 17-5. The new concrete should project slightly beyond the surface of the old concrete. It should be allowed to stiffen and then troweled and finished to match the adjoining surfaces.

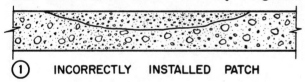

① INCORRECTLY INSTALLED PATCH

② CORRECTLY INSTALLED PATCH

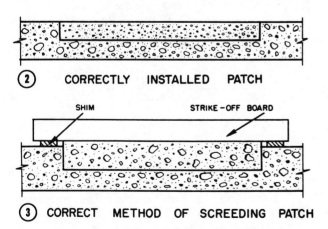

③ CORRECT METHOD OF SCREEDING PATCH

Patching Concrete
Figure 17-5

PATCHING OLD CONCRETE

Inspection. Before repairing old concrete, the amount of material to be removed must first be determined. A thorough inspection of the imperfection should be made before repairs are started. All concrete of questionable quality should be removed. However, in many cases, if all of the weakened material were removed, nothing would be left. In the event that all the weakened material cannot be removed and the old concrete will be completely encased in new concrete, only the loose material needs to be removed. Where old and new concrete form a junction at a surface exposed to weathering or chemical attack, the old concrete must be perfectly sound. It is far better to remove too much old concrete than too little.

Preparation. After initially removing the weakened material and loose particles, the surface to be repaired should be thoroughly cleaned with air or water, or both. The area around the repair should be kept continuously wet for several hours, preferably overnight. This wetting is especially important in the repair of old concrete. Without wetting a good bond cannot be achieved.

Patching Concrete. Where small areas of patching are involved, rectangular patches should be used. The upper 1 to 2 inches of the edge of the old concrete should be trimmed to a vertical face to eliminate the possibility of thin edges in the patch or in the old concrete. The depth of the repair is dependent upon many conditions. For repairing large structures such as walls, piers, curbs, and slabs, the depth of repair should be at least 6 inches where possible. If reinforcement bars are in the old concrete, there should be a clearance of at least an inch around each exposed bar. After the wetting period the new concrete should be placed in layers and each layer thoroughly tamped. The concrete should be a low-slump mixture which has been allowed to stand for a while in order to reduce shrinkage in the hole. In the repair of old concrete, it may be necessary to use forms to hold the new concrete in place. The design and construction of these forms often calls for a high degree of ingenuity. Well designed and properly constructed forms are important steps in the procedure for repairing concrete. Deep patches should be reinforced and to the hardened concrete. Following patching, good curing is essential. Curing should be started as soon as possible to avoid early drying.

CHAPTER 18

EFFECTS OF TEMPERATURE

HOT WEATHER CONCRETING

Problems. Concreting in hot weather poses some special problems, among which are reduction in strength and cracking of flat surfaces due to rapid drying. Concrete may stiffen before it can be consolidated because of rapid setting of the cement and excessive absorption and evaporation of mixing water. This leads to difficulty in the finishing of flat surfaces. During hot weather, precautions should be taken to limit concrete temperature to less than 90°. Limitations are imposed on the maximum temperature of concrete and on the placing of concrete during hot weather because the quality and durability suffer when concrete is mixed, placed, and cured at high temperatures. Difficulty can be experienced even with concrete temperatures of less than 90° F. The combination of hot, dry weather and high winds is most severe, especially when placing large exposed slabs.

Effects of High Concrete Temperatures.

Mixing water requirements. High temperatures accelerate the hardening of concrete and more mixing water is generally required for the same consistency. Figure 18-1 shows the increase in mixing water required to maintain the same slump as temperature increases; however, increasing the water content of concrete without increasing the cement content results in a higher water-cement ratio, thereby adversely affecting the strength and other properties of hardened concrete.

Compressive strength of concrete. Figure 18-2 shows the effect of high concrete temperatures on compressive strength. These tests, using identical concretes of the same water-cement ratio, show that while higher concrete temperatures increase early strength, at later ages the reverse is true. If the water content had been increased to maintain the same slump (without changing the cement content), the reduction in strength would have been even greater than shown in figure 18-2.

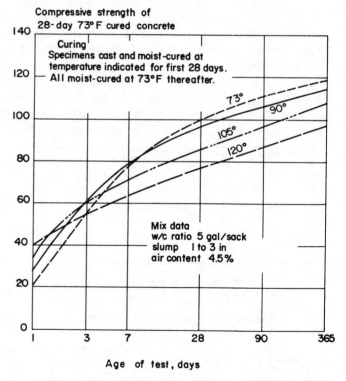

Effect Of High Temperature On Concrete
Compressive Strength At Various Ages
Figure 18-2

Cracking. In hot weather the tendency for cracks to form is increased both before and after hardening. Rapid evaporation of water from hot concrete may cause plastic shrinkage cracks

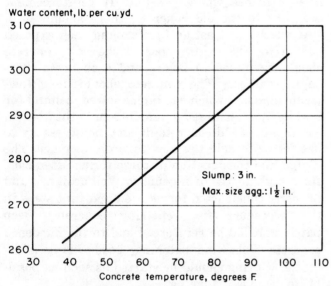

Increase In Water Requirement Of A Concrete
Mix As Temperature Increases
Figure 18-1

before the surface has hardened. Cracks may also develop in the hardened concrete because of increased shrinkage due to higher water requirement and because of the greater range between the high temperature at the time of hardening and the low temperature to which the concrete will later drop.

Cooling Concrete Materials. The most practical method of maintaining low concrete temperatures is to control the temperature of the concrete materials. One or more of the ingredients may be cooled before mixing. In hot weather the aggregates and water should be kept as cool as practicable. Mixing water is the easiest to cool and is the most effective, pound for pound, for lowering the temperature of concrete. However, since aggregates represent 60 to 80 percent of the total weight of concrete, the concrete temperature is primarily dependent on the aggregate temperature. Figure 18-3 shows the effect of the temperature of mixing water and aggregate on the temperature of fresh concrete. The temperature of the fresh concrete can be lowered by several means:

(1) Using cold mixing water. Slush ice can be used in extreme cases to cool the water.

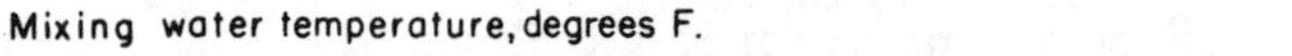

Mixing water temperature, degrees F.

Chart based on following mix proportions:
aggregate 3,000 lb.
moisture in aggregate 60 lb.
added mixing water 240 lb.
cement, at 150°F. 564 lb

Temperature Of Fresh Concrete As Affected
By Temperature Of Materials
Figure 18-3

(2) Cooling coarse aggregate by sprinkling; avoid excessive use of water.

(3) Insulating mixer drums or cooling them with sprays or wet burlap coverings.

(4) Insulating water-supply lines and tanks, or painting them white.

(5) Shading materials and facilities not otherwise protected from the heat.

(6) Working only at night.

(7) Avoiding the use of hot cement.

(8) Sprinkling forms, reinforcing steel, and subgrade with cool water just before placing concrete.

Additional Precautions. High temperatures increase the rate of concrete hardening and shorten the length of time within which the concrete can be handled and finished. Since setting is accelerated, transporting and placing should be done as quickly as practicable. Extra care must be taken in placing techniques to avoid cold joints. Curing is difficult in hot weather since the water evaporates rapidly from the concrete. Proper curing is especially important in hot weather because of the greater danger of crazing and cracking. Forms are not satisfactory substitutes for curing in hot weather. They should be loosened as soon as it can be done without damage to the concrete. Water should be applied and allowed to cover the concrete. The efficiency of curing compounds is reduced in hot weather. Frequent sprinkling and the use of wet burlap and other means of retaining the moisture for longer periods are necessary.

COLD WEATHER CONCRETING

Considerations. The placing of concrete does not have to be suspended during winter months if necessary precautions are taken to protect the fresh concrete from freezing temperatures until necessary protection can be provided. For successful winter work, adequate protection must be provided when temperatures of 40° F. or lower occur during placing and during the early curing period. Prior planning should include provisions for heating concrete materials and maintaining favorable temperatures after the concrete is placed. To prevent freezing, the temperature of the concrete should not be less than that shown in line 4, table 18-1, at the time of placing. Thermal protection may be required to assure that subsequent concrete temperatures do not fall below the minimum shown in line 5, table 18-1, for the periods shown in table 18-2, to ensure durability or to develop strength. The temperature of fresh concrete as mixed should not be less than shown in lines 1, 2, and 3 of table 18-1. Note that lower concrete temperatures are recommended for heavy mass sections than for thinner sections because less heat is dissipated during the hydration period. More heat is lost during transporting and placing; therefore, the fresh concrete temperatures are higher for colder weather. Temperatures for concrete over 70° F. are seldom necessary, since they do not furnish proportionately longer protection from freezing because the loss of heat is greater. High concrete temperatures also require more mixing water for the same slump; this contributes to cracking due to shrinkage.

Effect of Low Concrete Temperatures. Temperature affects the rate at which hydration of cement occurs—low temperatures retard concrete hardening and strength gain. This fact is illustrated in figure 18-4 for concrete mixed, placed and cured at temperatures between 40° and 73° F. It can be seen from the above figure that at temperatures below 73° F., the strength of concrete is lower during the first 28 days and then is higher than that of concrete cured at 73° F. Therefore, concrete placed at temperatures below 73° F. must be cured longer. It should be remembered that strength gain practically stops when moisture re-

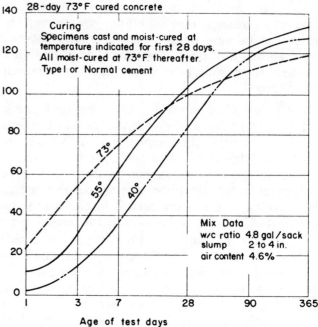

Effect Of Low Temperature On Concrete
Compressive Strength At Various Ages
Figure 18-4

quired for hydration is no longer available. The early strengths that may be achieved through use of Type III or High-Early Strength cement are higher than those achieved with Type I cement, as illustrated by figure 18-5.

Table 5-5. *Recommended Concrete Temperatures for Cold Weather Construction* (Air-Entrained Concrete)

Line	Condition of placement and curing		Thin sections	Moderate sections	Mass sections
1	Min. temp. fresh concrete *as mixed* for weather indicated, deg. F.	Above 30 deg. F.	60	55	50
2		0 to 30 deg. F.	65	60	55
3		Below 0 deg. F.	70	65	60
4	Min. temp. fresh concrete *as placed*, deg. F.		55	50	45
5	Max. allowable *gradual* drop in temp. throughout first 24 hours after end of protection, deg. F.		50	40	30

*Adapted from Recommended Practice for Cold Weather Concreting (ACI 306–66).

Recommended Concrete Temperatures For Cold Weather Construction* (Air-Entrained Concrete)
Table 18-1

Degree of exposure to freeze-thaw	Normal concrete**	High-early-strength concrete†
No exposure	2 days	1 day
Any exposure	3 days	2 days

**Made with Type I, II, or Normal cement.
†Made with Type III or High-Early-Strength cement, or an accelerator, or an extra 100 lb. of cement.

Recommended Duration Of Protection For Concrete
Placed In Cold Weather* (Air-Entrained Concrete)
Table 18-2

Cold Weather Techniques.

Heating concrete materials. Thawing frozen aggregates makes proper batching easier. Frozen aggregates must be thawed to avoid pockets of aggregate in the concrete after placing. Excessively high water content must be avoided if thawing takes place in the mixer. It is seldom necessary to heat aggregates in temperatures above freezing. At temperatures below freezing, con-

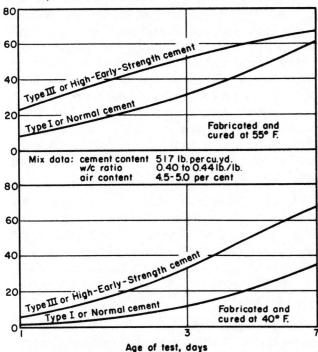

Compressive strength, per cent of 28-day 73° F. cured concrete

Fabricated and cured at 55° F.

Mix data: cement content 517 lb. per cu. yd.
w/c ratio 0.40 to 0.44 lb./lb.
air content 4.5-5.0 per cent

Fabricated and cured at 40° F.

Age of test, days

Early Compressive Strength Relationships Involving
mmmPortland Cement Types And Low Curing
Temperature
Figure 18-5

crete of the required temperatures may be produced by heating the fine aggregate only.

Heating aggregates. Several methods are used to heat aggregates. On small jobs aggregates may be heated by piling them over metal pipes in which fires are built. To obtain recommended concrete temperatures, the average temperature of the aggregates should not exceed 150° F. Aggregates are frequently stockpiled over pipes through which steam is circulated. The stockpiles should be covered with tarpaulins to retain and distribute the heat. Live steam may be injected directly into the pile of aggregate to heat it but the problem of variable moisture content can result in erratic control of mixing water.

Heating water. Mixing water is the easiest ingredient of concrete to heat. It can store five times as much heat as solid materials of the same weight, although the weight of aggregates and cement is much greater than the weight of water. The heat stored in water may be used to heat concrete materials. When either aggregates or water are heated above 100° F., they should be combined in the mixer before the cement is added. Figure 18-6 shows the effect of the temperature of materials on the temperature of fresh concrete.

Figure 18-6 is reasonably accurate for most ordinary concrete mixtures. Water should not be hotter than 180° F. shown on the chart. The chart further indicates that in some cases both aggregates and water must be heated. For example, if the weighted average temperature of aggregates is below 36° F. and the desired concrete temperature is 70° F., then the aggregates would have to be heated in order to limit the water temperature to 180° F.

Mixing water temperature degrees F

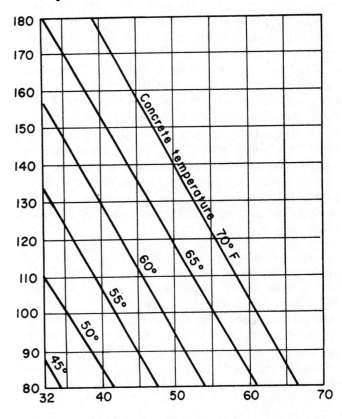

Concrete temperature 70° F

Weighted average temperature of aggregates and cement, degrees F

Chart based on following mix proportions

aggregate	3,000 lb
moisture in aggregate	60 lb
added mixing water	240 lb
cement	564 lb

Temperature Of Mixing Water Needed To Produce
Heated Concrete Of Required Temperature
Figure 18-6

Use of high-early-strength concrete. High strength at an early age is frequently desired during winter construction to reduce the length of time protection is required. The value of high-early-strength concrete during cold weather is often realized through early re-use of forms and removal of shores, savings in the cost of additional heating and protection, earlier finishing of flatwork, and earlier use of the structure.

Use of accelerators. So called antifreeze compounds or other materials should not be used to lower the freezing points of concrete. Strength of concrete and other properties are seriously affected by the large quantities of an accelerator required to lower the freezing point appreciably. Small amounts of additional cement or of accelerators such as calcium chloride to accelerate hardening of concrete in cold weather may be beneficial, as long as no more than 2 percent of calcium chloride by weight of cement is used. Precautions are necessary when using accelerators containing chlorides where there is an in-service potential for corrosion, as, for example, for prestressed concrete, or where aluminum inserts are contemplated. When sulfate-resisting concrete is required, use an extra sack of cement per cubic yard rather than calcium chloride. Accelerators should not be used as a substitute for proper curing and frost protection.

Preparation before placing. Concrete should never be placed on a frozen subgrade. Severe cracks usually occur due to settlement when the subgrade thaws. If the subgrade is frozen for only a few inches deep, the surface may be thawed by burning straw, by steaming, or if the grade permits, by spreading a layer of hot sand or other granular material. The ground must be thawed enough to assure that it will not freeze during the curing period.

Curing. Concrete in forms or covered with insulation seldom loses enough moisture at 40° to 55°F. to impair curing. Some moisture must be provided for concrete curing during winter to offset the drying tendency when heated enclosures are used. Concrete should be kept at a favorable temperature until it is strong enough to withstand low temperatures and anticipated service loads. Forms serve to distribute the heat more evenly and help prevent drying and overheating. They should be left in place as long as practicable. Concrete that is allowed to freeze soon after placing is permanently damaged. If the concrete has been frozen once at an early age, it may be restored to nearly normal under favorable curing conditions although it will not weather as well nor be as watertight as concrete that is not frozen. Air-entrained concrete is less susceptible to damage from freezing than concrete without entrained air.

(1) *Live steam.* Live steam exhausted into an enclosure is an excellent practical method of curing in extremely cold weather because moisture from the steam offsets the rapid drying that occurs when very cold air is heated. A curing compound may be used after the protection is removed and the air temperature is above freezing.

(2) *Insulation blanket or bat insulation.* The manufacturers of these materials can usually provide information on the amount of insulation necessary for protection at various temperatures. The corners and edges of concrete are most vulnerable to freezing and should be checked to determine the effectiveness of the protective covering.

(3) *Heated enclosures.* Wood, canvas, building board, plastic sheets, or other materials are used to enclose and protect concrete from below freezing temperatures. Wood framework covered with tarpaulins or plastic sheets is also used. The enclosures should be sturdy and reasonably airtight. Free circulation of warm air should be provided for. Control of the temperature within the enclosure is easiest with live steam. Carbon-dioxide-producing heaters (salamanders and other fuel-burning heaters) should not be used during concrete placing and for 24 to 36 hours after placement unless they are properly vented. Temperature differences should be minimized. Adequate minimum temperatures should be provided for the entire curing period.

CHAPTER 19

REINFORCED CONCRETE CONSTRUCTION

Reinforced Concrete. Concrete is strong in compression, but relatively weak in tension. The reverse is true for slender steel bars and when the two materials are used together one makes up for the deficiency of the other. When steel is embedded in concrete in a manner which assists it in carrying imposed loads, the combination is known as reinforced concrete. Beam strength can be increased significantly by the use of steel in the tension side (figure 19-1).

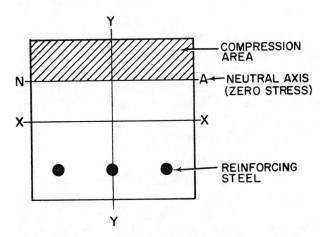

Cross-Section Of A Reinforced Concrete Beam
Figure 19-1

CHARACTERISTICS OF CONCRETE

Tensile Strength. Tensile strength is such a small percentage of the compressive strength that it is ignored in reinforced concrete beam calculations. Instead, tensile resistance is provided by longitudinal steel bars well embedded in the tension side.

Shear Strength. The shear strength of concrete is about one-third the unit compressive strength, and tensile strength is less than one-half the shear strength. A concrete slab subjected to a downward concentrated load fails due to the diagonal tension. Beams are prevented from failing in diagonal tension by providing web reinforcement.

Bond Strength. Bond strength is the resistance developed by concrete to the pulling out of a steel bar embedded therein. The theory of reinforced concrete beam design is based on the as-

sumption that a bond exists between the steel and concrete which prevents relative movement between them as the load is applied. The amount of bond strength that can be developed depends largely upon the area of contact between the two materials. Due to their superior bond value, bars manufactured with a very rough outside surface, called deformed bars (figure 19-2), have replaced plain bars.

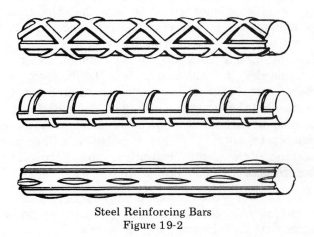

Steel Reinforcing Bars
Figure 19-2

Bending Strength. When a beam is subjected to a bending moment it deflects because the parts

I. REINFORCEMENT

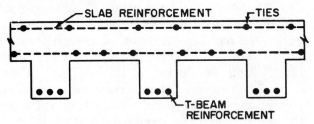

2. EFFECTS OF VERTICAL LOAD

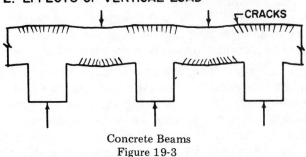

Concrete Beams
Figure 19-3

that are in compression shorten and those that are in tension become longer. The weak portions of the beam, as shown by the short irregular lines in figure 19-3 where tension exists must be reinforced with steel. This figure is not intended to show that beams always crack excessively, but to show the condition that beams may reach if they are loaded sufficiently. Concrete in the areas subjected to compression is usually effective by itself.

Creep. The tendency for loaded concrete to deform after a lapse of time is known as creep or plastic flow. Concrete tends to exhibit this continuing deformation over the whole stress range. It takes place rapidly at first, then much more slowly, becoming small or negligible after a year or two. Some authorities believe that the initial strains of well-designed reinforced concrete structures are removed during the first few service loadings and after that they perform elastically as long as they are not overloaded. Due to creep, deflection of concrete cannot be predicted by the common deflection formulas with any satisfactory degree of accuracy. However, failures are not usually traceable to creep because this phenomenon usually ceases in well proportioned structures before excessive deflections occur.

Homogeneous Beams. Beams composed of the same material throughout, such as steel or timber beams are called homogeneous beams.

Neutral Axis. The axis where the bending stress in a beam is zero is called the neutral axis (figure 19-1).

REINFORCED CONCRETE DESIGN

Specifications. The solutions to common problems of reinforced concrete design are influenced by the practical experience of many structural engineers and by the results of exhaustive tests and investigations conducted at universities and elsewhere. The results of these experiences, tests, and investigations have been reflected in rules and methods that are published as the "Report of the Joint Committee on Standard Specifications for Concrete and Reinforced Concrete" and other references. In most practical designs engineers will make reference to standard specifications.

Design. The term "design of a beam" for instance denotes determination of the size and the materials required to contract a beam that can safely support specified loads under certain definite conditions of span, stress, and the like. Economy and efficiency in the use of materials, strength, spacing, and arrangement of reinforcing steel are factors which enter into the design. The design of a reinforced concrete structure requires sound engineering judgment and experience. No attempt is made to teach the design of reinforced concrete members or structures in this manual in view of the many authoritative texts available in this field. The design of reinforced concrete consists principally in predicting the position and direction of potential tension cracks in concrete, and in forestalling the cracking by locating sufficient steel across them. From the structural engineer's point of view, there are three types of members, namely, tension members, compression members and bending members, called beams. Beams require the most study because the bending stress varies over the cross section, instead of being uniformly distributed.

Workmanship. It is emphasized that the best of designs can be ruined if the intent of the plans is not carried out faithfully and intelligently in the field. Proper reinforced concrete construction depends on men who understand the action of structures and who appreciate the characteristics and limitations of the material.

STRUCTURAL MEMBERS

Types of Structural Members. A reinforced concrete structure is made up of many types of reinforced structural members including columns, beams, girders, walls, footings, slabs, etc. Analysis indicates that the different members interact to a considerable degree in view of the fact that a reinforced concrete structure is monolithic.

Beam Reinforcement. Four common types of beam reinforcing steel are shown in figure 19-4. Both straight and bent-up principal reinforcing bars are depended on to resist the bending tension in the bottom over the central portion of the span. Fewer bars are necessary on the bottom near the ends of the span where the bending moment is small. For this reason, some bars may be bent as shown in figure 19-4 so the inclined portion can be used to resist diagonal tension. The reinforcing bars of continuous beams are continued across the supports to resist tension in the top in that area.

(1) When there are not enough bent bars available to resist all the diagonal tension, additional U-shaped bars, called stirrups, are usually necessary. Due to the tensile stress on the stirrups, they must pass under the bottom steel and

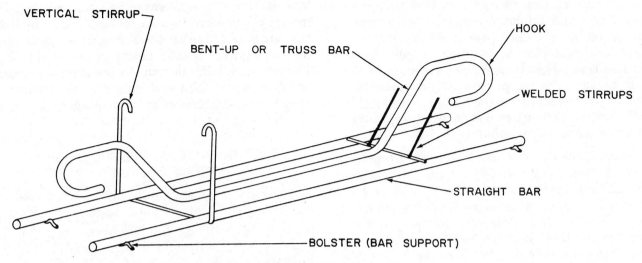

Typical Shapes Of Reinforcing Steel
Figure 19-4

perpendicular to it to prevent lateral slippage. Welded stirrups serve the same purpose as U-shaped bars and may be placed at any desired angle decided on by the engineer.

(2) Horizontal reinforcing steel is usually supporting on devices called bolsters or chairs (figure 19-5) that hold the bars in place during construction. Stress-carrying reinforcing bars must be placed in accordance with American Concrete Institute (ACI) Code 318–63, section 808, Building Code Requirements for Reinforced Concrete.

Column Reinforcement. A column is a slender, vertical member which carries a superimposed load. Concrete columns must always be reinforced with steel, unless the height is less than three times the least lateral dimension in which

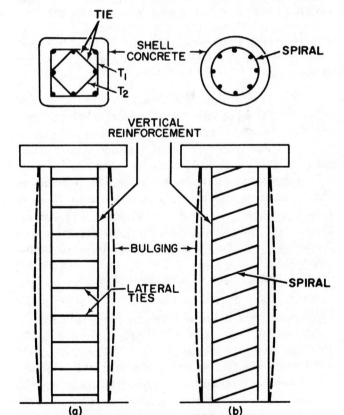

Reinforcing Concrete Columns
Figure 19-6

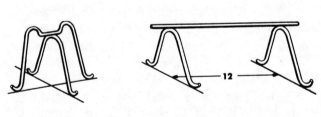

Supports For Reinforcing Steel
Figure 19-5

case the member is called a pier or pedestal. Allowable loads and minimum column dimensions are governed by ACI Code 318–63. Most concrete columns are subjected to bending. Figure 19-6 shows two types of column reinforcement. Vertical reinforcement is the principal reinforcement. Lateral reinforcement surrounds the column horizontally and consists of individual ties ((a), figure 19-6) or a continuous spiral ((b), figure 19-6).

Tied Columns. A loaded concrete column shortens vertically and expands laterally ((a), figure 19-6). Lateral reinforcement in the form of lateral ties is used to restrain the expansion. The principal value of lateral reinforcement is to provide intermediate lateral support for the vertical, or longitudinal reinforcement. Columns reinforced in this manner are called tied columns.

Spiral Columns. A spiral column is identified by a continuous spiral winding ((b), figure 19-6) which encircles the core and longitudinal steel. A spiral column is generally considered to be more substantial than a tied column due to the continuity of the spiral reinforcement, as opposed to the many imperfect anchorages at the ends of the individual lateral ties in a tied column. The pitch of the spiral reinforcement can be reduced to provide effective lateral support. The pitch of the spiral and tie size and number of bars are specified by the engineer.

Composite and Combination Columns. A structural steel or cast iron column thoroughly encased in concrete, reinforced with both longitudinal and spiral reinforcement is called a composite column. The cross-sectional area of the metal core of a composite column cannot exceed 20 percent of the gross area of the column. A structural steel column encased in concrete at least $2\frac{1}{2}$ inches thick over all the metal and reinforced with welded wire fabric is called a combination column. Composite and combination columns are often used in construction of large buildings.

Vertical Reinforcement. The vertical reinforcement in a column helps to carry the direct axial load as the column shortens under load. Vertical bars are located around the periphery of a column for effective resistance to possible bending. Each vertical reinforcing bar tends to buckle outward in the direction of least opposition. For this reason every vertical reinforcing bar should be held securely, at close vertical intervals against outward lateral movement. For example in the 8-bar group shown in (b), figure 19-6, a second sys-

tem of ties, T_2 is necessary to confine the four intermediate vertical reinforcing bars. If these ties are omitted, the 8-bar group tends to come into a slightly circular configuration under load. The resultant bulging leads to destructive cracking of the concrete shell and failure of the column. A round column has obvious advantages in this respect.

SHRINKAGE AND TEMPERATURE REINFORCEMENT

Slabs and walls must not only be reinforced by the principal reinforcement against the applied loads to which they are subjected. They must also be reinforced in the lateral direction to resist the effects of shrinkage and temperature change. Concrete shrinks as hydration proceeds. A small percentage of steel must be used to resist this force. Similarly, concrete must be allowed to contract with the lowering of temperature. A fall in temperature of about 83 degrees causes as much movement as drying shrinkage. Depending on the extent adjacent construction interferes, this movement tends to cause the concrete to become stressed in tension. The amount of shrinkage and temperature reinforcement usually provided for is approximately one-fifth of 1 percent of the area of the cross section of concrete, as required by the specifications.

REINFORCING STEEL

Grades. Concrete reinforcing steel is available as bars in 11 bar sizes (table 19-1) ranging from $\frac{3}{8}$-inch diameter to about $2\frac{1}{4}$-inch diameter, as wire for column spirals, and as wire mesh which is often used for temperature and shrinkage reinforcements of slabs and walls. Reinforcing steel is specified in ASTM A615, A616, and A617. The minimum yield strengths are: 40,000 psi, 50,000 psi, 60,000 psi, and 75,000 psi. The engineer will specify the size, amount and classification of reinforcing steel to be used. Wire mesh is cold-drawn from hard-grade steel. All steel used in the manufacture of concrete reinforcing bars is ductile. Generally, hooks of relatively small radius can be cold-formed on the job without breakage. Improved deformed bars, which conform to ASTM Specifications are so superior in bond value that hooking the ends of the bar does not add significantly to the strength. Bars which do not meet ASTM specifications are no longer rolled for concrete reinforcement. With so many grades of steel now available, quite likely having the same pattern of deformation, some permanent, rolled-on identification is necessary and is required by the latest ASTM specifications. Two systems of grade markings have been adopted and are illustrated in figure 19-7.

(1) Continuous longitudinal smaller lines between the main longitudinal ribs. One small line identifies 60,000 psi; two such lines 75,000 psi yield strength.

(2) Rolled-on numbers follow the symbols and number that indicates bar size. The number 40 identifies 40,000 psi; the number 50 identifies 50,000 psi; the number 60 identifies 60,000 psi, and the number 75 identifies 75,000 psi yield strength steel.

CONTINUOUS LINE SYSTEM-GRADE MARKS

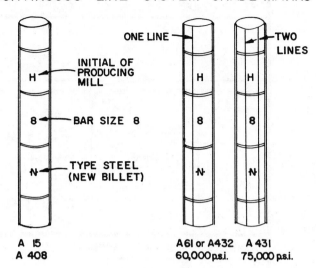

NUMBER SYSTEM - GRADE MARKS

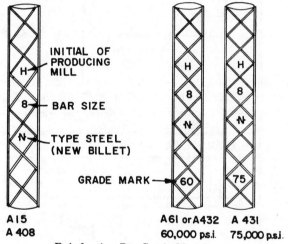

Reinforcing Bar Grade Markings
Figure 19-7

Bar designation No.**	Unit weight lb/ft.	Diameter in.	Cross-sectional area, in.²	Perimeter in.
3	.376	0.375	0.11	1.178
4	.668	0.500	0.20	1.571
5	1.043	0.625	0.31	1.963
6	1.502	0.750	0.44	2.356
7	2.044	0.875	0.60	2.749
8	2.670	1.000	0.79	3.142
9	3.400	1.128	1.00	3.544
10	4.303	1.270	1.27	3.990
11	5.313	1.410	1.56	4.430
14	7.65	1.693	2.25	5.32
18	13.60	2.257	4.00	7.09

*The nominal dimensions of a deformed bar are equivalent to those of a plain round bar having the same weight per foot as the deformed bar.
**Bar numbers are based on the number of eighths of an inch included in the nominal diameter of the bars.

Standard Steel Reinforcing Bars*
Table 19-1

Hooks. It is not possible or feasible in some cases to extend straight bars far enough to develop their strength by bond only. In these cases, it is common practice to specify bent or hooked rods to obtain the required length for their development through bond. The hook provides some mechanical locking of the steel into the concrete. Good engineering judgment will frequently dictate means to insure suitable anchorage for the end of the rod, particularly if the bar has to resist large tension very near the end. Hooks are usually necessary for reinforcing bars in tension unless the bars terminate at the tops of continuous beams. When it is practicable, the longitudinal rods in a beam are anchored in the compression area. Details for bending hooks in reinforcing bars are shown in table 19-2. The 90° hook may be used

when a little mechanical anchorage is desired but full development of the rod is not necessary. As indicated, a reasonably large radius is required for this type of hook. The 135° hook is less desirable than the 90° hook because the acute angle of the bend tends to increase the compression in the concrete at the inside of the bend. It is usually preferable to bend the rod 180° to withstand the compression caused by tension in the steel. The straight portion beyond the bend is also desirable as additional anchorage. All bends in reinforcing bars should have reasonably large diameters so that the full length of the rod can be considered to be effective. A hook should not be used to anchor a bar subjected to compression.

RECOMMENDED SIZES – 180° HOOK

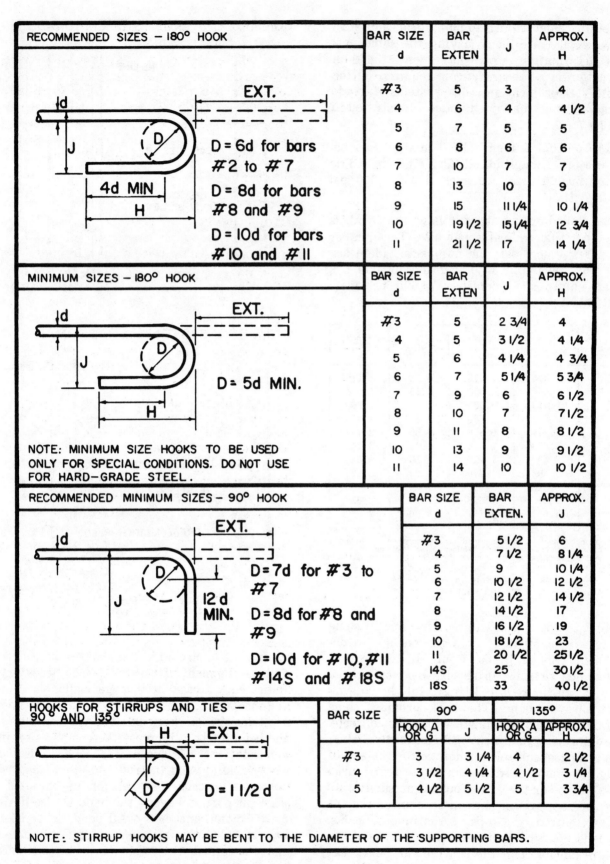

D = 6d for bars #2 to #7

D = 8d for bars #8 and #9

D = 10d for bars #10 and #11

BAR SIZE d	BAR EXTEN	J	APPROX. H
#3	5	3	4
4	6	4	4 1/2
5	7	5	5
6	8	6	6
7	10	7	7
8	13	10	9
9	15	11 1/4	10 1/4
10	19 1/2	15 1/4	12 3/4
11	21 1/2	17	14 1/4

MINIMUM SIZES – 180° HOOK

D = 5d MIN.

BAR SIZE d	BAR EXTEN	J	APPROX. H
#3	5	2 3/4	4
4	5	3 1/2	4 1/4
5	6	4 1/4	4 3/4
6	7	5 1/4	5 3/4
7	9	6	6 1/2
8	10	7	7 1/2
9	11	8	8 1/2
10	13	9	9 1/2
11	14	10	10 1/2

NOTE: MINIMUM SIZE HOOKS TO BE USED ONLY FOR SPECIAL CONDITIONS. DO NOT USE FOR HARD-GRADE STEEL.

RECOMMENDED MINIMUM SIZES – 90° HOOK

D = 7d for #3 to #7

D = 8d for #8 and #9

D = 10d for #10, #11 #14S and #18S

12 d MIN.

BAR SIZE d	BAR EXTEN.	APPROX. J
#3	5 1/2	6
4	7 1/2	8 1/4
5	9	10 1/4
6	10 1/2	12 1/2
7	12 1/2	14 1/2
8	14 1/2	17
9	16 1/2	19
10	18 1/2	23
11	20 1/2	25 1/2
14S	25	30 1/2
18S	33	40 1/2

HOOKS FOR STIRRUPS AND TIES – 90° AND 135°

D = 1 1/2 d

BAR SIZE d	90°		135°	
	HOOK A OR G	J	HOOK A OR G	APPROX. H
#3	3	3 1/4	4	2 1/2
4	3 1/2	4 1/4	4 1/2	3 1/4
5	4 1/2	5 1/2	5	3 3/4

NOTE: STIRRUP HOOKS MAY BE BENT TO THE DIAMETER OF THE SUPPORTING BARS.

Hook Details For Reinforcing Steel
Table 19-2

Bolsters and Chairs. Bolsters or chairs (fig. 19-8) are available in a variety of heights to suit any situation. Bolsters or chairs keep the bars carrying computed stress from 1 to 2 inches from the outside shell. This is frequently called protective concrete.

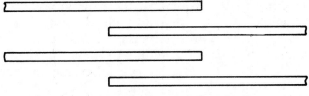

1. BARS LAPPED IN MIDDLE OF SPANS

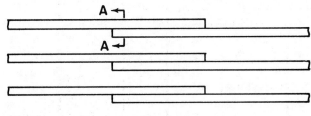

ENLARGED SECTION A-A

2. BARS LAPPED SIDE BY SIDE IN HORIZONTAL PLANE

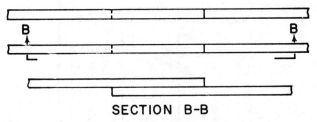

SECTION B-B

3. BARS LAPPED IN VERTICAL PLANE
NOTE: SPLICES SHOULD BE STAGGERED

Reinforcing Bar Splices
Figure 19-8

Stirrups. Small U-shaped bars called stirrups are used to supplement bent bars in resisting diagonal tension and to reinforce the web in a beam to prevent cracks from spreading. They must pass underneath the bottom steel and be perpendicular to it to prevent lateral slippage. Vertical stirrups can be readily arranged and so easily set in the forms with the other rods that they are one of the most practical systems of web reinforcement. Anchorage of the stirrups is secured by welding to longitudinal steel, by hooking tightly around longitudinal reinforcement, and by embedding sufficiently above the mid-depth of the beam to develop the required stress by bond. Welded stirrups (fig. 19-4) may be placed at any angle. They serve the same purpose as vertical stirrups in resisting diagonal tension in the beam. Inclined stirrups are more efficient than vertical ones since they may be oriented parallel with diagonal stress. Some practical considerations offset this advantage. This type of stirrup should be welded to longitudinal reinforcement to avoid slippage and displacement during concrete placement, but this work is expensive and somewhat troublesome. Stirrups must be placed so that every potential diagonal tension crack is crossed by at least one stirrup.

SPLICES

Methods. The usual method of splicing reinforcing bars is by lapping the bars past each other so that bond stress will transfer the load from one bar into the concrete and then into the other bar (figure 19-8). The rods could be hooked but it is not always practicable or even desirable to bend them. The length of the lap is a matter of engineering judgment in considering the stresses anticipated in the beam, but approximates 24 to 36 rod diameters, depending on the size of the rod. For plain bars the minimum length of the lap shall be twice that for deformed bars. Splices should not be made at the points of maximum bending. It is usually best to locate splices beyond the center of the beam. When possible, splices should be staggered so that all the splices do not come at the same point. The method shown in 1, figure 19-8 is satisfactory when the spacing of the bars is large but it is undesirable in a beam or similar member having several closely spaced bars when the overlapped section sometimes interferes with proper encasement of rods and the filling of forms.

Types. The lap in a horizontal plane illustrated by 2, figure 19-8 is the most practical arrangement if the spacing provides enough clearance for the passage of aggregate. Both this method and the one shown in 3, figure 19-8 facilitate tying the bars to hold them in position during concreting. However, the tying does not add significantly to the strength of the splices. There is a possibility of air pockets and poor bond in the space under the junction between the bars. Lapping of rods in a vertical plane as shown in 3, figure 19-8 has the advantage of better encasement but the top rods do not fit the stirrups properly, and the beam has a smaller effective depth at one place than at another one. In practice, bars in this position are likely to be knocked down in the position shown in 2, figure 19-8. Reinforcing bars are not butted when one rod is not long enough for the span. *Except as shown on plans, no splicing of reinforcement should be made without approval of an engineer.* There should be at least two supports under every bar and bolsters should be spaced at about 5-foot intervals.

Storing. Excessive rusting of the reinforcing steel in storage should be avoided. Before the steel is placed, the surface should be free from objectionable coatings, particularly heavy corrosion caused by outdoor storage. When stored outside, reinforcing bars should be placed on dunnage.

Cleaning. A thin film of rust or mill scale is not considered to be seriously objectionable—in fact, it may increase the bond of steel with concrete, but loose rust or scale which can be removed by rubbing with burlap or by other means should be removed. Other objectionable coatings commonly found on reinforcing steel are oil, paint, grease, dried mud, and weak dried mortar. If the mortar is difficult to remove, it will probably do no

harm where it is. Anything that destroys the ability of the concrete to grip the steel may prove to be serious if it prevents the stress in the steel from performing its function properly.

Fabrication. When large numbers of reinforcing bars of varied lengths and shapes are required they can be prefabricated on the job according to the drawings. Stirrups and column ties are usually less than $\frac{1}{2}$ inch in diameter and can be bent cold. Steel bars larger than $\frac{3}{8}$ inch in diameter to be used for main reinforcement should be cold-bent. Heating is normally unnecessary except for bars over $1\frac{1}{8}$ inch in diameter. The bends, except for hooks, should be made around pins having a diameter of not less than six times the bar diameter. If the bar is larger than 1 inch the minimum diameter of the bending pin should be eight times the bar size. Steel bars larger than $\frac{3}{8}$ inch in diameter should be bent with a bar bending machine whenever possible. If a bending machine is not available, a hickey or the bar bending table shown in figure 19-9 may be used, although the bends made with these devices are usually too sharp and the bar is weakened. A hickey, a level for bending bars, may be impro-

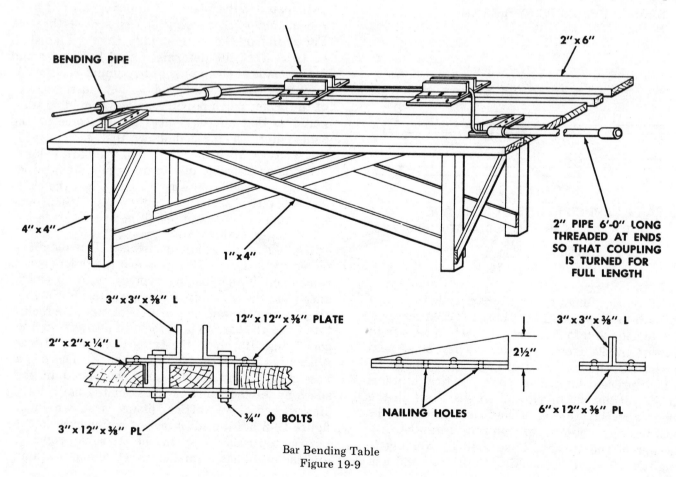

Bar Bending Table
Figure 19-9

vised by attaching a 2- by 1½- by 2-inch pipe tee to the end of a 1¼-inch pipe lever 3 feet long, and sawing a section from one side of the tee.

PLACEMENT

Support. All steel reinforcement should be accurately located in the forms and firmly held in place before and during the casting of concrete. This can be done by means of built-in concrete blocks, metallic supports, spacer bars, wires, or other devices adequate to insure against displacement during construction and to keep the steel at the proper distance from the forms. There should be enough supports and spacers to carry the steel properly even when subject to construction loads. The use of rocks, wood blocks or other unapproved objects to support the reinforcing steel is prohibited. Horizontal bars should be supported at minimum intervals of 5 or 6 feet. All bars should be secured to supports and to other bars by tie wires. Wire used to tie bars should not be smaller than 18 gage. The twisted ends of ties should project away from an interior surface.

Spacers. Some specifications require that no metal be left in the concrete within a given distance of the surface. Generally, the minimum clear distance between parallel bars in beams, footings, walls, and floor slabs should not be less than 1⅓ times the largest size aggregate particle in the concrete nor less than 1 inch. Spacer and supporting blocks can be made from a mortar with the same consistency as the concrete, but without the coarse aggregate. The spacer blocks are usually 1½ inches square or larger varying in length as required. Tie wires are cast in the blocks to secure the blocks to the reinforcing bars. When this type of spacer block is used, removal is not necessary when the concrete is placed.

Columns. The clear distance between parallel bars in columns should not be less than 1½ times the bar diameter. The steel for columns is first tied together and placed in position as a unit.

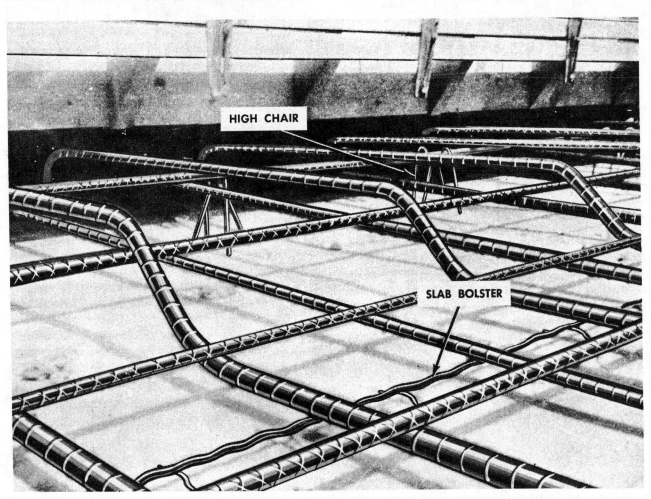

Reinforcing Steel For A Floor Slab
Figure 19-10

Then the column form is erected around the unit and the reinforcing steel is tied to the form at 5-foot intervals.

Floor Slabs. A typical arrangement of reinforcing steel in floor slabs is shown in figure 19-10. The height of the slab bolster is determined by the thickness of the required concrete protective cover. Concrete blocks made of sand-cement mortar can be used in place of the slab bolster. The bars should be tied together with one turn of wire at frequent intervals where they cross, to hold them firmly in place.

Beams. Reinforcing steel for a reinforced concrete beam shown in figure 19-11, indicates the position of bolsters and stirrups. Note that the stirrups pass under the main reinforcing rods and are tied to them with one turn of wire.

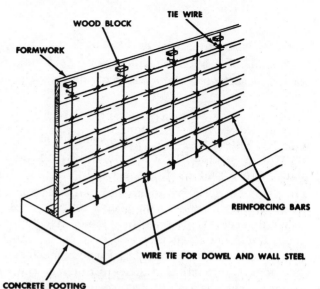

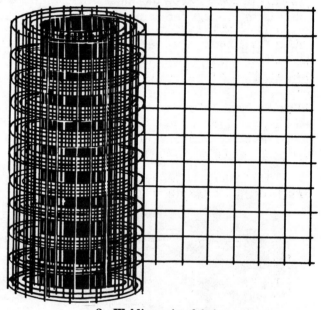

1 Reinforcing bars for a wall

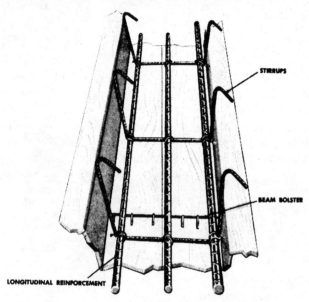

Beam Reinforcing Steel
Figure 19-11

Walls and Footings. Reinforcing steel is erected in place for walls, unless wire fabric is used. It is not preassembled as it is for columns. Ties between the top and bottom should be used for high walls. The wood blocks (1, figure 19-12) are for high walls. The wood blocks (1, figure 19-12) are removed when the forms have been filled up to the level of the block. Welded wire fabric (2, figure 19-12) is also used as concrete reinforcement for footing, walls and slabs. Reinforcing steel for footings should be placed after the forms have been set. A typical arrangement is shown in 3, figure 19-12. Concrete bars or clean sound stones may be used in footings to support the steel the proper distance above the subgrade.

2 Welding wire fabric

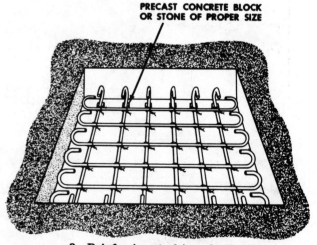

3 Reinforcing steel in a footing
Wall And Footing Reinforcement
Figure 19-12

150

CHAPTER 20

PRECAST CONCRETE

Definition. Precast concrete is any concrete member that is cast in forms at a place other than its final position of use. The member may be of either plain or reinforced concrete. It can be done anywhere although this procedure is best adapted to a factory or yard. Job-site precasting is not uncommon for large projects. Some manufacturers produce a variety of structural members in several different shapes and sizes, including piles, girders, roof members and other standard products. Prestressed concrete is especially well adapted to precasting techniques.

Products. Generally, structural members including standard highway girders, piles, electric poles, masts and building members are precast by factory methods unless the difficulty or impracticability of transportation makes job-site casting more desirable. The economies obtained by precasting standard members or members required in large numbers in a central location, for a particular project, are readily apparent.

Advantages. Economy of mass production is the principal advantage of precasting. Added to

this is the desirability of fabricating on the ground rather than in the final position the member is intended for.

Disadvantages. Some of the inherent advantages of precasting are offset by the necessity for extensive plant facilities including equipment storage space. In addition, some advantages are offset by the cost of transporting and the necessity for heavy equipment to place precast members in position.

Transporting Precast Members. Prestressed members can be hauled further and given rougher treatment without detriment than either plain or reinforced members. In all cases, care must be taken to support the members in such a way that no excessive strain or loading is applied that is different from the design loading. Various types of hauling and handling equipment are used in precast concrete operations. Heavy girders to be transported long distances can be hauled with tractor trailers or with a tractor and dolly arrangement in which the girder acts as the tongue or tie between the tractor and the dolly. Smaller

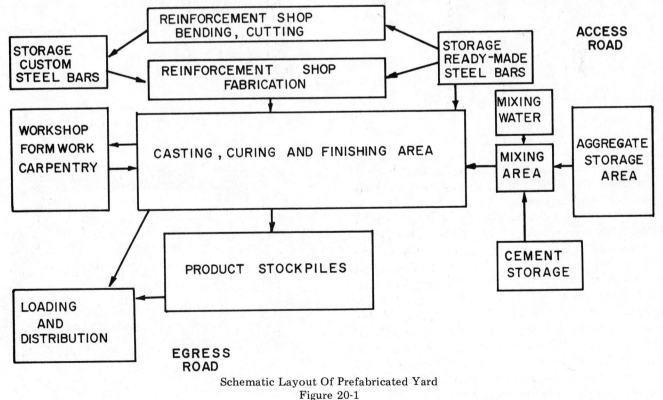

Schematic Layout Of Prefabricated Yard
Figure 20-1

151

members such as columns, piles, slabs, etc., can be hauled on flatbed trailers. These members must be protected from excessive bending stresses due to their own dead weight.

Erecting Precast Elements. The erection of precast members is similar to steel erection. Cranes or derricks of sufficient capacity are the usual means of lifting the members. Spreader bars frequently must be used in order to handle the elements at the correct pickup points. In some cases, due to long length or heavy weight, two cranes or derricks may be necessary to lift the members.

PREFABRICATION YARD

Precasting is done either in central prefabrication plants or at site prefabrication plants depending upon the product and its application. On site or temporary prefabrication plants are generally more suitable for typical operations. These plants are without roofing and therefore subject to weather and climate considerations. The prefabrication yard is laid out to suit the type and quantity of members to be processed. It must be on firm level ground providing ample working space and access routes. Bridge T-beams, reinforced concrete arches, end walls, and concrete logs are typical members produced at these plants. A schematic layout of a prefabrication yard suitable for producing such members is shown in figure 20-1. A prefabrication unit of this size can be expected to produce approximately 6000 square feet of precast walls per day. The output will vary according to personnel experience, equipment capabilities, and product requirements.

CHAPTER 21

CLEANING CONCRETE AND MASONRY

METHODS

General. Materials to be cleaned and the type of spots or stains to be removed determine the cleaning process used. Sandblasting and acid-washing work well on some brick, concrete, granite, and other hard surfaces, but are harmful to soft stone surfaces and glazed finishes. A method that cleans marble or glazed surfaces successfully may be entirely unsatisfactory on concrete or rough-textured clay-brick surfaces. Safe methods for a masonry surface may damage mortar joints between masonry units. If the surface has more than one kind of masonry, one cleaning method may not be safe or effective on all.

Concrete. Many chemicals can be applied to concrete and concrete masonry without much injury, but chemicals having an acid reaction should not be used. Even weak acids may roughen the surface if left on it long. Stains on concrete and masonry caused by iron, copper, bronze, aluminum, ink, tobacco, urine, fire, lubricating oil, rotted wood, coffee, and iodine, and general service stains can usually be removed by the methods listed below. Old, long-neglected stains may require repeated treatments. Remove deep stains with a poultice or bandage treatment. Make the poultice by mixing active chemicals with fine inert powders to a pasty consistency and apply a thick layer to the stained surface. Prepare the bandage by soaking cotton batting or several layers of cloth in chemicals; paste bandage over the stain. More detailed methods of stain removal are described later in this chapter.

Masonry. Masonry units of low absorption and those with smooth or glazed surfaces generally respond readily to proper cleaning methods and resume their original appearance. Highly absorbent or rough-surfaced units are more difficult to clean. If the stain has penetrated the pores of the unit, cleaning may remove part of the surface, destroy its texture, and change its appearance. Principal methods of cleaning masonry structures are with steam, water, sandblasting, and various liquids and pastes.

Scaffolds. Use either a painter's overhead, suspended type or stationary ground-supported scaffold when cleaning surfaces high above ground. The kind used depends on building height and other local conditions.

STEAM CLEANING

Cleaning with high-pressure steam and water, sometimes called cold-steam cleaning, is effective and economical. It removes grime from weather-exposed surfaces of concrete and masonry buildings without harming surfaces or preventing later weathering. After steam cleaning, masonry surfaces retain their original finish and natural color tones without the roughened surfaces and dulled arrises caused by sandblasting or the dead, bleached tone caused by acid cleaning. Steam cleaning equipment including boiler must be inspected and certified by qualified personnel prior to start of operations. Provisions of the ASME Boiler Code must be complied with. Proposed cleaning and operating procedure must be carefully evaluated to assure continued safe operating conditions.

Equipment. Use proper equipment for steam-cleaning. Much of it can be improvised, but local requirements may make buying or renting equipment advantageous.

Steam supply. See that there is a continuous supply of high-pressure steam or water vapor before cleaning operations begin. A portable boiler with accessories, truck-mounted for easy transportation, is generally a satisfactory supply source. Steam pressure for cleaning old buildings is preferably 150 pounds and *never* less than 140 pounds. For cleaning new work, 120 pounds, *never* less than 100 pounds, are needed. Boiler capacity of about 12 horsepower for each cleaning nozzle is necessary. The pressure with which the steam and water mixture is driven against the wall, not volume of discharge, is important in steam cleaning.

Nozzle. The cleaning nozzle is a most important accessory. It should be a mixing type having a water-control valve and automatic steam shut-off. Several nozzles are available. One efficient type has a very narrow opening 4 inches long and can deliver an extremely fine spray at a high velocity. Operating two cleaning nozzles from each length of scaffold is good practice.

Water-supply hose. Ordinary garden hose is suitable for carrying water from source of supply to scaffold. Use only high-pressure steam couplings, with suitable pipe and fittings, if pipe is used to convey steam from boiler to scaffold.

Rinsing hose. In addition to the hose supplying water to the steam-cleaning nozzle, have another hose (ordinary garden-type nozzle with shut-off) to flush the walls with water occasionally.

Procedure.

Cleaning. Cleaning is done by high-velocity projection of a finely divided spray of steam and water against the masonry surface. The mixture of steam and water spray entering minute surface depressions and openings dissolves and dislodges grime, soot, and other extraneous matter, which is later flushed down the wall by the rinsing hose. Experimentation with equipment used quickly shows the operator the best angle and distance of nozzle from the wall and proper regulation of steam and water valves for most effective work. Work on one 3-foot-square space at a time. Pass the nozzle back and forth over the area, then flush with clear water before moving to the next space.

Cleaning additives. Alkalies such as sodium carbonate, sodium bicarbonate, and trisodium phosphate are sometimes added to the cleaning water to speed the cleaning action. While these salts aid in cleaning, some are retained in the masonry, and may appear on the wall later as efflorescence. To cut down the amount of salts retained, wet wall units thoroughly with clear water before beginning the cleaning operations. Immediately after cleaning, wash the wall with plenty of clear water to remove all possible salts from the face of the wall.

Removing Stains. Removal of surface dirt sometimes reveals stains. A mild acid wash may be necessary to remove them. After the stain is removed, steam the treated surface again and flush with water from the rinsing hose to remove all trace of acid wash.

Removing Hardened Deposits. Steel scrapers or wire brushes may be necessary to remove hardened deposits that cannot be removed by steam cleaning. Use scrapers sparingly and carefully to avoid damage

to masonry surface. Use brushes of fine spring-steel wire to grind off the hard deposit without digging into or scratching the cleaned surface. After the hard deposit has been removed by scraping or wire brushing, flush off the surface with water, treat it with the steam-cleaning nozzle, and flush again to remove all loose dirt which might cause future streaking or discoloration.

SANDBLAST CLEANING

Sandblasting cleans rust and scale from structural-steel members, bridges, gas holders, oil tanks, and many other metal surfaces efficiently. However, do not use this method on marble, terra cotta, glass or units with glazed or other special surfaces or textures. Although it cleans effectively, it often destroys the original surface of the masonry unit. It tends to dull sharp edges, blur ornamental detail and carving, and roughen surfaces. Stone cleaned by sandblasting, especially limestone, appears whiter, but this whiteness is caused partly because the stone surface has been bruised. In sandblasting, compressed air forces hard sand through a nozzle against the masonry surface. The sand removes accumulated grime and a thin layer of masonry. For best results follow the procedures listed below.

Hose. Use a ¾-inch hose if maximum pressure of air and sand is required to remove dirt, and a 1-inch hose if volume rather than high pressure is desired.

Sand. Quality of sand needed varies with depth of cutting done. Placing a canvas screen around the scaffold platform keeps sand from scattering and makes it possible to salvage about 75 percent of the sand.

Personnel. Usually, four men operate a sandblasting outfit; one at the machine, one at the nozzle, and two on the ground to handle hose, scaffold, sand, and so on.

Repointing Mortar Joints. When hardness between masonry units and mortar joints differs widely, sandblasting may cut too deeply into the mortar. If this occurs, joints must be repointed.

Waterproofing. Because roughened surfaces produced by sandblast cleaning quickly gather soot and dirt, an application of transparent waterproofing is desirable. This coating fills surface pores, tends to make the wall self-cleaning, and prevents rapid soiling of the surfaces by smoke and dust.

154

ACIDS, CAUSTIC WASHES AND PASTE CLEANERS

Selection of Cleaner. Acid, caustic concentrations, and paste mixtures are used to clean many interior and exterior concrete masonry surfaces. The cleaning material used depends on the type of surface to be cleaned, age of structure, and material or stain to be removed. The most satisfactory results can be obtained by developing a cleaning material for the job at hand after analyzing the dirt and stains to be removed. Strength and chemical composition of the cleaner can usually be adjusted by observing the action on trial spaces. Protect grass and shrubbery from damage.

Cautions in Using Acids.

Do not use acid solutions to clean limestone and similar materials unless experienced operators and expert supervision are available. Acid washes tend to eat into the stone surfaces and pit them. They usually bleach, producing an unnatural appearance, and may cause yellow stains to appear later. If an acid solution is not thoroughly washed from the masonry pores after cleaning, the destructive action continues for some time.

When mixing acid solutions, always pour acid into water.

Handle acid solutions carefully because they are harmful to the skin and especially to the eyes. When working with acid cleaners, wear goggles, gloves, and protective clothing, and keep a supply of running water at hand.

Use only wooden containers and fiber brushes when cleaning with acids. Do not use wire brushes or steel wool to scrub the walls, because small steel particles become lodged in crevices, producing rust spots and stains.

Soap Powder. Many burned-clay and concrete surfaces and glazed and polished surfaces of tile, marble, and glass can be cleaned by hand-scrubbing with a white soap powder dissolved in soft water, using ordinary fiber scrub brushes. Wash surfaces thoroughly with clear water after scrubbing them.

Mortar Stains. To clean mortar stains, first remove excess mortar with a putty knife or a chisel. Soak the wall with clear water and wash it by applying one of the following acid solutions with a fiber brush or broom:

Ten parts water to 1 part hydrochloric (muriatic) acid Federal Specification O–H–765, technical grade, 31 percent solution.

Twenty parts water to one part phosphoric acid, commercial grade, 75 percent solution.

Since acid solutions attack mortar joints, wash the wall thoroughly with clear water immediately after it is cleaned. Remove final traces of acid by applying a dilute ammonia solution (1 pint of ammonia to 2 gallons of water).

Efflorescence. Water applied with stiff scrubbing brushes frequently removes efflorescence. If this does not remove all stains, follow the procedures described above, using a water-hydrochloric acid solution.

Stains on Cement Stucco. To clean white cement stucco, wet the surface, apply a solution of 20 parts water and 1 part sulfuric acid (commercial grade, 93 percent solution) and rinse thoroughly with clear water. Hydrochloric acid solutions may produce a yellowish tinge on white cement.

Paints and Similar Coatings.

Whitewash, calcimine, and cold-water paints. To remove whitewash, calcimine, and cold-water paints, wash the surface with an acid solution of one part muriatic acid and five parts water. Scrub vigorously with a fiber brush as the solution forms. When the coating has been removed, wash the wall with water from an open hose until all trace of acid is removed.

Oil paints, enamels, varnishes, shellacs, and glue sizings.

Remove oil paint, enamel, varnish, shellac, or glue sizing by applying a paint remover, leaving it on until softened paint can be scraped off with a putty knife, or flushed off with water. After using paint remover, wash the wall thoroughly to remove all traces of acid. Efficient paint removers include—

1. An acceptable noncombustible, low toxicity paint remover.
2. Two pounds trisodium phosphate in 1 gallon hot water.
3. Two and ½ pounds caustic soda in 1 gallon hot water.
4. One part sodium hydroxide dissolved in three parts water and added to one part mineral oil. Stir mixture until emulsified, then stir in one part sawdust or other inert material.
5. Equal parts soda ash and quicklime mixed with enough water to form a thick paste. Leave this mixture on the wall for 24 hours after application,

then scrape off.

If the oil-paint film is very thick and hard, sandblasting may be the best way to remove it. Burning is sometimes used, but is not recommended because of the fire hazard.

If a paint film is old, crumbling, and flaking, scraping it off with wire brushes and metal scrapers may be necessary. While this method is effective, it may leave metal particles which later cause rust stains in the wall surface.

CLEANING OF MISCELLANEOUS STAINS

Use one or more of the following materials and methods to remove stains from clay, concrete masonry, stone and marble:

Iron Stains. Iron stains can usually be recognized by their resemblance to iron rust or by their proximity to steel or iron members in the building. Large areas of concrete or cement stucco may be stained if curing water used contains iron. Remove by mopping the surface with a solution of 1 pound of oxalic acid dissolved in 1 gallon of water. After 2 or 3 hours, rinse with clean water, scrubbing at the same time with stiff brushes or brooms. Some spots may require a second mopping and scrubbing. For older, deeper stains the following methods are recommended:

Method. 1. Dissolve one part sodium citrate in six parts lukewarm water. Mix thoroughly with seven parts of lime-free glycerine. Add to this solution enough whiting or kieselguhr to make a paste poultice stiff enough to adhere to the surface when applied with a putty knife or trowel to a thickness of ¼ inch or more. Allow a minimum of 2 days for drying. Scrape off and wash thoroughly. If the stain has not disappeared, repeat the treatment. This treatment has no injurious effects, but its action may be too slow for bad stains. Ammonium citrate produces quicker results, but may injure a polished surface slightly, making a repolish job necessary.

Method 2. The sodium hydrosulphite combination bandage-poultice method is more satisfactory for removing deep, intense, iron stains.

(*a*) Make a solution by dissolving one part sodium citrate crystals in six parts of water. Dip white cloth or cotton batting in this solution, place the cloth over the stain, and leave it there for 15 minutes.

(*b*) On horizontal surfaces, sprinkle a thin layer of hydrosulphite crystals over the stain being treated with sodium citrate, moisten with water, and cover with a paste of whiting and water.

(*c*) Give vertical surfaces the sodium-citrate treatment. Place a layer of whiting paste on a plasterer's trowel, sprinkle on a layer of hydrosulphite crystals, moisten slightly, and apply it to the stain. Remove treatment after 1 hour. If it is left on longer, a black stain may develop. Wash treated surface with clean water. If inspection shows incomplete removal of the iron stain, repeat the cleaning operation, using fresh materials.

Tobacco Stains.

Method 1. Dissolve 2 pounds of trisodium-phosphate crystals in 4 to 5 quarts of hot water. Mix 12 ounces of chlorinated lime to a smooth stiff paste in a shallow enameled pan by adding water slowly and mashing the lumps. Pour this and the trisodium-phosphate solution into a 2-gallon stoneware jar and fill it with water. Stir well, cover the jar, and allow lime to settle. Add some of the liquid to powdered talc, stirring until a thick paste is obtained. Apply the paste with a trowel to a poultice thickness of ¼ inch. Scrape off the dry paste with a wooden paddle or trowel. This mixture is a strong bleaching and corrosive agent. Care must be taken not to drop it on colored fabrics or metal fixtures.

Method 2. Minor tobacco stains can be removed by applying ½-inch-thick poultices of a stiff paste made of water and any of the several grit scrubbing powders commonly used on marble, terazzo, and tile floors. Scrape off the dry paste with a wooden paddle and wash the surface with clean water. In most cases, two or more applications will be necessary.

Fire and Smoke Stains.

Method 1. Fire and smoke stains can sometimes be removed by scouring with powdered pumice or a grit scrubbing powder. After removing the surface stain by scouring, the deep-seated stains can be removed by applying the trisodium-phosphate-chlorinated-lime solution described above for tobacco stains. Fold a white canton flannel cloth to three or four thicknesses and saturate it with the liquid. Paste this saturated cloth over the stain and cover it with a slab of concrete or sheet of glass, making sure the cloth is pressed firmly against the stained surface. If the surface is vertical, devise a method to hold the saturated cloth firmly against the stain. Resaturate the cloth from time to time. Wash the surface thoroughly at the end of the treatment.

Method 2. Make a smooth stiff paste of trichlorethylene and powdered talc, and apply it as a troweled-on poultice. Cover the poultice with glass or a pan to prevent rapid evaporation. Allow time to dry. Scrape off and wash away all traces of treatment material. Trichlorethylene gives off harmful fumes, therefore, see that closed spaces are well ventilated when using this stain remover.

Copper and Bronze Stains. These stains are nearly always green, but in some cases may be brown. Mix one part dry amonium chloride (sal ammoniac) and four parts powdered talc, add water, and stir to a thick paste. Trowel ¼-inch layer of paste over the stain and leave until dry. When working on polished marble or similar fine surfaces, use a wooden paddle to scrape off dried paste. An old stain may require several applications. Aluminum chloride may be used instead of sal ammoniac.

Aluminum Stains. These appear as a white deposit which can be removed by scrubbing with a 10 to 20 percent muriatic acid solution. On colored concrete, use the weaker solution. Wash thoroughly with clean water.

Oil Stains. Oils penetrate most concrete readily. Oil spilled on horizontal surfaces should be covered immediately with a dry powdered material such as hydrated lime, Fuller's earth, or whiting. Sweep up the powdered material, taking as much of the oil as possible. Scrub with a solution of 1.1.1 trichlorethane, technical, inhibited (methyl chloroform) Federal Specifications O–T–620. If the treatment is made soon enough, there will be no stain. However, when oil has remained for some time, one of the following methods may be necessary:

Method 1. Mix 1 pound of trisodium phosphate in 1 gallon of water and add sufficient whiting to make a stiff paste. Spread a layer ½ inch thick over the surface to be cleaned. Leave paste until it dries (about 24 hours), remove, and wash with clear water.

Method 2. Saturate white canton flannel in 1.1.1 trichloroethane, technical, inhibited (methyl chloroform) Federal Specification O–T–620, and place it over the stain. Cover the cloth with a slab of dry concrete or sheet of glass. If stain is on a vertical surface, improvise means to hold cloth and covering in place. Keep the cloth saturated until the stain is removed. Covering saturated cloth with glass tends to drive the stain in, while the slab of dry concrete will draw out some of the oil.

Coffee Stains.

Saturate a cloth in a solution of one part of glycerin and four parts of water. Place the cloth over the stain and resaturate the cloth from time to time until the stain disappears.

Javelle water (sodium hypochlorite) can be used as the bleaching agent instead of the glycerin and water solution. Javelle water can usually be purchased at drug stores. If not, it is prepared as follows: Dissolve 3 pounds of common washing soda in 1 gallon of water. Mix 12 ounces of chlorinated lime to a paste in a shallow enameled iron pan by adding water slowly and mashing the lumps. Add the paste to the soda solution, pour the mixture into a stoneware jar, and add water to 2 gallons. Stir thoroughly, cover jar, and allow lime to settle. Draw off the clear liquid and dilute it with six times its volume of clear water. Use as a soap or scrubbing solution, rinsing the surface thoroughly before and after application. Javelle water is a strong bleaching material and should not be allowed to drop on colored fabrics. It is not recommended for general cleaning purposes.

Iodine Stains. Iodine stains gradually disappear of their own accord. Hasten removal of stains from horizontal surfaces by applying alcohol covered with whiting or talcum powder. If stain is on a vertical surface, make a paste of talcum and alcohol, brush alcohol on the stain, and cover with the paste. Allow paste to dry, scrape it off and wash surface with clear water.

Perspiration Stains. Secretions from the hands and oils from the hair produce brown or yellow stains that may be mistaken for iron stains. Removal methods described for fire and smoke stains are recommended. Bad stains may require several treatments.

Urine Stain. Use the method described for tobacco stains. Should the stain prove stubborn, saturate cotton batting in one of the liquids, paste it over the stain, and resaturate it from time to time until the stain is removed. Complete the treatment by washing with clear water.

Ink Stains. Stains made by different inks may require dissimilar removal treatments.

(1) The acid content of many ordinary inks generally causes an etching action on concrete, masonry, and marble surfaces. Prompt removal should be attempted. Make a strong solution of sodium perborate in hot water. Mix with enough whiting to make a thick paste, apply a ¼-inch layer over the stain, and leave until dry. If some blue color remains after poultice is removed, repeat the treatment. If a brown stain remains, treat it by the method for iron stains.

(2) Many bright-colored inks are water solutions of synthetic dyes. The sodium-perborate poultice generally removes these stains. Other removal treatments are—

Cover stain with a cotton-batting bandage saturated with ammonia water.

Use Javelle water in the same way as ammonia water, or mix it with whiting to make a thick paste and apply a layer over the stain.

Use as a poultice a mixture of chlorinated lime and whiting reduced to paste and water.

(3) Some blue inks containing prussian blue, or ferrocyanide of iron, produce a stain which cannot be removed by the perborate or chlorinated lime poultices, or the Javelle-water treatments. Such stains can be treated by covering them with cotton batting saturated with ammonia water. Strong soap solutions applied in the same way may also produce satisfactory results.

(4) Indelible inks often consist entirely of synthetic dyes. These stains may be treated as described in (1) above. Some indelible inks may contain silver salts which produce black stains. These stains can be removed by applying bandages saturated with ammonia water. Usually several bandage applications are necessary.

APPENDIX A

METHOD OF MAKING SLUMP TEST FOR CONSISTENCY OF PORTLAND CEMENT CONCRETE

1. SCOPE

This method of testing covers the procedure to be used in the laboratory and in the field for determining consistency of concrete. It is not an exact method but gives sufficiently accurate results.

This test is not considered applicable when there is a considerable amount of aggregate over 2 inches in size in the concrete.

2. APPARATUS

The following apparatus is required and is available in the mobile laboratory.

A mold of No. 16 gauge galvanized metal in the form of the lateral surface of the frustum of a cone with the base 8 inches in diameter, the top 4 inches in diameter, and the altitude 12 inches. The base and the top are open and parallel to each other and at right angles to the axis of the cone (figure A-1).

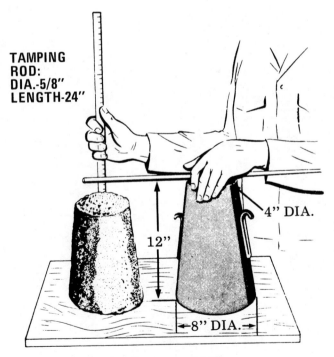

TAMPING ROD: DIA.-5/8" LENGTH-24"

4" DIA.

12"

8" DIA.

Measurement Of Slumps
Figure A-1

3. SAMPLES

Samples of concrete for test specimens will be taken at the mixer or, in the case of ready-mixed concrete, from the transportation vehicle during discharge. The sample of concrete from which test specimens are made will be representative of the entire batch. Such samples will be obtained by repeatedly passing a scoop or pail through the discharging stream of concrete, starting the sampling operation at the beginning of discharge and repeating the operation until the entire batch is discharged. The sample thus obtained will be transported to the testing site. To counteract segregation the concrete shall be mixed with a shovel until the concrete is uniform in appearance. The location in the work of the batch of concrete thus sampled will be noted for future reference. In the case of paving concrete, samples may be taken from the batch immediately after depositing on the subgrade. At least five samples will be taken from different portions of the pile and these samples will be thoroughly mixed to form the test specimen.

4. PROCEDURE

The mold will be dampened and placed on a flat, moist non-absorbent surface. From the sample of concrete obtained as described in paragraph 3, the mold will immediately be filled in three layers, each approximately one-third the volume of the mold. In placing each scoopful of concrete the scoop will be moved around the edge of the mold as the concrete slides from it, in order to insure symmetrical distribution of concrete within the mold. Each layer will be rodded with 25 strokes of a 5/8-inch round rod, approximately 24 inches in length and tapered for a distance of 1 inch to a spherically shaped end having a radius of approximately 1/4 inch. The strokes will be distributed uniformly over the cross section of the mold and will penetrate into the underlying layer. The bottom layer will be rodded throughout its depth. After the top layer has been rodded, the surface of the concrete will be struck off with a trowel so that the mold is exactly filled. The mold will be immediately removed from the concrete by raising it carefully in a vertical direction. The slump will then be measured immediately by determining the difference between the height of the mold and the height at the vertical axis of the specimen.

5. SLUMP

The consistency will be recorded in terms of inches of subsidence of the specimen during the test, which shall be known as the slump--

Slump == 12 inches minus height of specimen after subsidence.

6. SUPPLEMENTARY TEST PROCEDURE

After the slump measurement is completed, the side of the specimen frustrum should be tapped gently with the tamping rod. The behavior of the concrete under this treatment is a valuable indication of the cohesiveness, workability, and placeability of the mix. A well-proportioned workable mix will gradually slump to lower elevations and retain its original identity, while a poor mix will crumble, segregate, and fall apart.

APPENDIX B

ESTIMATING QUANTITIES AND LABOR HOURS FOR CONCRETE, FORMS AND REINFORCING

Concrete is a simple material to take off because the shapes are regular and are usually clearly shown on plans. Sometimes the footings are an exception and are shown schematically. They should be checked carefully in the take off. Piers are often shown undersize and are dimensioned in a table on the plans.

In Figure B-1 there is represented a cross-section of a typical foundation wall and footings. If the wall is L feet long the following formulas apply:

Volume of footing $= WtL$

Volume of wall $= MhL$

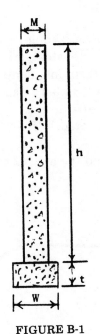

FIGURE B-1

The proper method of take-off is to sum up all lengths of the same cross-section and make one volume calculation for each section. For example if a footing 1'-6" by 8" is used for lengths of footing of 12', 18', 24' and 22', the proper method to be used for calculating the volume is to add: 12 +

18 + 24 + 22. The result, 76' is used to find the volume.

$$Vol. = 1.5' \times 0.67' \times 76' = 76.4 \text{ CF}$$

In order to simplify the calculation, use is often made of tables which contain the volume of concrete for various footing cross-sections one foot long. Table B-1 is such a table. Walls are often taken off by the square foot of surface because then one calculation can serve for both concrete and formwork. Table B-2 gives the cubic feet of concrete required for 100 square feet of wall.

VOLUME OF CONCRETE IN FOOTINGS PER FOOT OF LENGTH		
Width (Ft.)	Depth (Ft.)	Volume Per Foot of Length (Cu. Ft./Ft.)
1'-0"	6"	0.50
1'-2"	6"	0.585
1'-4"	6"	0.665
1'-6"	6"	0.75
1'-0"	8"	0.67
1'-2"	8"	0.784
1'-4"	8"	0.889
1'-6"	8"	1.00
1'-0"	10"	0.833
1'-2"	10"	0.972
1'-4"	10"	1.11
1'-6"	10"	1.25
1'-8"	10"	1.39
1'-10"	10"	1.53
2'-0"	10"	1.67
1'-2"	1'-0"	1.17
1'-4"	1'-0"	1.33
1'-6"	1'-0"	1.5
1'-8"	1'-0"	1.67
1'-10"	1'-0"	1.83
2'-0"	1'-0"	2.00

TABLE B-1

CUBIC FEET OF CONCRETE PER 100 SQUARE FEET OF WALL	
Wall Thickness (In.)	Concrete (Cu. Ft. /100 Sq. Ft.)
4	33.3
5	41.7
6	50.0
7	58.3
8	66.7
9	75.0
10	83.3
11	91.7
12	100.0

TABLE B-2

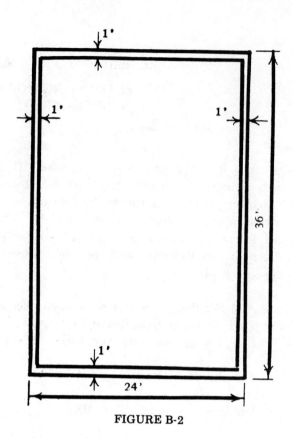

FIGURE B-2

In taking off footings and walls, care should be employed so that the corners are not counted twice when measuring lengths. In Figure B-2 we would take off:

$$2 @ 36' \qquad = 72'$$
$$2 @ (24'-2') \quad = \underline{44'}$$
$$\qquad\qquad\qquad 116'$$

The 2 feet were subtracted from the 24 foot length because this part of the wall had already been measured in the 36 feet length.

Wall and footing quantities are taken off by combining all equal cross sectional area and multiplying by the total length for each section. Footings and walls are taken off in cubic feet.

They may or may not be changed into cubic yards depending upon the practice of the particular contractor. Small jobs are usually calculated in cubic feet while larger jobs are more often calculated in cubic yards.

Sidewalks and slabwork are taken off by measuring the total area and multiplying by the thickness. Where sections are simple and the principal dimensions are written on the plans, the area is calculated by employing the simple geometrical formulas. However when the areas are irregular it is often easier to divide them into simpler forms by drawing straight lines on the plan. The required dimensions are then scaled off the smaller areas and the individual areas calculated. Then all areas are added.

Example: Find the area of the paved area shown in Figure B-3.

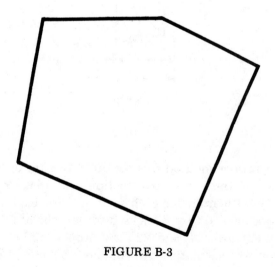

FIGURE B-3

First divide the area into simpler areas as shown in Figure B-4.

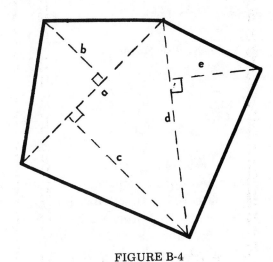

FIGURE B-4

$$\text{Areas} = \frac{ab}{2} + \frac{ac}{2} + \frac{ed}{2}$$

The volume is calculated by multiplying the total area in square feet by the required depth of slab in decimals of a foot.

Before following the method just given check some given dimension of the area of interest to be certain that the area has been drawn to scale.

Reinforced concrete when of simple conventional shapes is often measured in cubic yards or feet and the steel included in the unit price per volume of reinforced concrete. Some contractors on the other hand always separate the concrete and steel no matter how small the job. For large jobs and other than simple walls and footings, most contractors will separate the concrete and steel quantities.

REINFORCING

Reinforcing is estimated by the pound for small quantities or ton for large quantities. See Table B-3.

All bars of the same diameter are added and multiplied by the appropriate weight per foot. The total of the weights of all bars are then extended by a unit cost per pound or ton.

SIZES AND WEIGHTS OF REINFORCING BARS		
BAR NUMBER	SIZE (IN..)	WEIGHT (Lb./Ft.)
2	1/4 round	0.167
3	3/8 round	0.376
4	1/2 round	0.668
5	5/8 round	1.043
6	3/4 round	1.502
7	7/8 round	2.044
8	1 round	2.670
9	1 square	3.400
10	1 1/8 square	4.303
11	1 1/4 square	5.313

TABLE B-3

Reinforcing bars are shown on the plans by bar numbers. The relationship existing among bar numbers, sizes and weights is shown in Table B-3. Bar lengths should be measured by tabulating the lengths shown on the plans including the specified allowance for lap and bending. It is helpful to use a form similar to that shown in Table B-4 for reinforcing take off.

Retaining walls are often detailed as shown in Figure B-5. In this case the size of each bar, its spacing and often the number and length are found in tables on the plan. The take off is made in the manner just described. All concrete take off is simple once it is broken down into its basic parts.

Wire mesh reinforcing may be purchased in widths up to 8 feet. Estimate the area to be covered in square feet and include all areas to within one inch of the slab edge. Allow 10% for lap where widths exceed 8 feet.

Bar Number	Length (Feet)	Number of bars			Hooks	Bends
		0'–10'	10'–20'	20'–30'		

REINFORCING BAR SCHEDULE

TABLE B-4

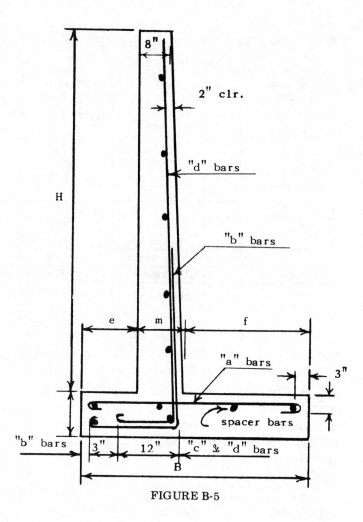

FIGURE B-5

DIMENSIONS AND WEIGHT OF WELDED WIRE FABRIC

Spacing	Wire Gauge	Pounds Per 100 SF
4" x 12"	No. 6 x No. 6	41.6
4" x 12"	No. 8 x No. 8	29.6
6" x 12"	No. 4 x No. 4	43.8
6" x 12"	No. 6 x No. 6	31.8
6" x 6"	No. 6 x No. 6	42.0
6" x 6"	No. 8 x No. 8	30.0
6" x 6"	No. 9 x No. 9	25.0
6" x 6"	No. 10 x No. 10	20.7
4" x 4"	No. 6 x No. 6	61.9
4" x 4"	No. 8 x No. 8	44.1

ESTIMATING LABOR HOURS FOR CONCRETE REINFORCING

Items	Hours
Placing reinforcing bars (no ties), hours per 100 lbs.	1.0
Placing reinforcing bars (tied in place) hours per 100 lbs.	1.3
Placing wire mesh, hours per 100 S.F.	.5

TABLE B-5

FORM MATERIAL

Side forms will not be required for footings except where soil conditions are such that the banks are not self-supporting. Forms will be used for both sides of all foundation walls. Experience indicates that for foundation walls up to 8 feet in height, approximately 2 board feet of lumber will be required for forms for each square foot of contact surface.

Example: Estimate the square feet of forms and board feet of lumber for a poured concrete wall 40 feet long x 6 feet high. Since forms will be required for both sides, the square feet of forms (contact area) will be obtained as follows:

SF of forms   = 40 x 6 x 2   = 480 SF
BF of lumber = 2 x 480   = 960 BF
The BF of lumber would be broken down as follows:
    1" boards 480 (contact area) plus 17 percent waste = 560 BF
    2 x 4's 960 - 560 = 400 BF

Material used for forms will often be used later in the structure as studding, bridging, sheathing, or

LABOR HOURS FOR ERECTING AND REMOVING WOOD CONCRETE FORMS

Item	Hours Per 100 SF Contact Area	
	Carpenter	Helper
Foundation Walls	5	4
Columns And Piers	10	7.5
Steps	9	6

TABLE B-6

subflooring. Consequently it is generally not necessary to include material costs when estimating costs of concrete forms.

ESTIMATING LABOR HOURS FOR PLACING AND FINISHING CONCRETE

Concrete footings and foundation walls, per cubic yard, unskilled labor	Hours	
Ready mixed concrete, labor placing	2.0	
Job hand-mixed, mixing and placing	4.0	
Job machine mixed, mixing and placing	3.0	

Concrete floor slabs, per 100 SF, unskilled labor	3" Thick	4" Thick
Ready mixed, concrete, labor placing	2.0	2.5
Job hand mixed, mixing and placing	4.0	5.0
Job machine mixed, mixing and placing	3.0	3.7

Cement Finishing, per 100 SF	Skilled Hours	Unskilled Hours
Troweling directly on slab with cement mortar topping	1.2	1.2
1/4" topping, mixing, spreading and troweling	1.2	1.5
1/2" topping, mixing, spreading and troweling		1.7

TABLE B-7

Amount of Mixing Water for Concrete

Proportions			Mixing Water Required Per Bag		Mixing Water Required Per Cubic Yard	
Cement	Sand	Stone	Minimum (gallons)	Maximum (gallons)	Minimum (gallons)	Maximum (gallons)
1	1½	3	5½	6	42	46
1	2	3	5¾	6¼	40	43½
1	2	4	6	6½	36	39
1	2½	5	7¼	7¾	36	38½
1	3	6	8¼	8¾	35	37

TABLE B-9

AGGREGATE PROPORTIONS FOR VARIOUS TYPES OF CONCRETE WORK

These mixes are suggested as suitable under normal or average conditions. If pebbles or crushed rock are not available the following cement and sand mixes may be substituted:

1:2½ mix for 1:2:3 mix; 1:3 for 1:2:4 mix; 1:3½ mix for 1:2½:4 mix.

Work	Mixture	Maximum Size aggregate	Slump	Water Cu Ft. per Sack	Ratio Gals. per Sack
Areaways	1:2½:4	1½ in.	2½ in.	.80	6.00
Barn Approaches	1:2½:4	1½ in.	2½ in.	.80	6.00
Bins	1:2:3	1½ in.	4 to 6 in.	.65	4.50
Boiler Settings	1:2:4	2 in.	4 to 6 in.	.70	5.25
Catch Basins	1:2:3	1½ in.	2 to 3 in.	.60	4.50
Cellars	1:2:4	1½ in.	4 to 6 in.	.75	5.50
Cisterns	1:2:3	1 in.	4 to 6 in.	.65	4.50
Cold Frames	1:2½:4	1 in.	4 to 6 in.	.90	6.75
Courts, Tennis	1:2½:4	1½ in.	2 to 4 in.	.85	6.25
Curbs	1:2½:4	1½ in.	2 to 4 in.	.85	6.25
Dipping Vats	1:2:3½	1½ in.	4 to 6 in.	.70	5.25
Driveways	1:2:4	1½ in.	2 to 4 in.	.70	5.25
Engine Beds	1:2:4	2 in.	4 to 6 in.	.75	5.50
Fence Posts	1:2:3	½ in.	2 to 4 in.	.60	4.50
Floors, Reinforced	1:2:3	1 in.	4 to 6 in.	65	4.50
Floors, Plain	1:2½:4	1½ in.	2 to 4 in.	.85	6.25
Foundations (Mass)	1:2½ 5	3 in.	4 to 6 in.	.90	6.75
Gutters	1:2½:4	1½ in.	2 to 4 in.	.85	6.25
Hog Wallows	1:2:3½	2 in.	2 to 4 in.	.65	4.50
Hot Beds	1:2½:4	1 in.	4 to 6 in.	.90	6.75
Manure Pits	1:2:3	1½ in.	4 to 6 in.	.65	4.50
Piers, House	1:2:3½	2 in.	4 to 6 in.	.70	5.25
Retaining Walls	1:2:3½	1½ in.	4 to 6 in.	.70	5.25
Roads	1:2:3½	2 in.	1 to 3 in.	.65	4.50
Roofs	1:2:3	1½ in.	4 to 6 in.	.65	4.50
Runways	1:2½:4	1½ in.	2 to 4 in.	.80	6.00
Sidewalks	1:2½:4	1½ in.	2 to 4 in.	.80	6.00
Steps and Stairways	1:2½:4	1 in.	2 to 4 in.	.80	6.00
Slabs	1:2:3	1½ in.	4 to 6 in.	.65	4.50
Septic Tanks	1:2:4	1 in.	4 to 6 in.	.75	5.50
Storage Cellar Walls	1:2½:4	1½ in.	4 to 6 in.	.85	6.25
Stucco	1:4	¼ in.	1 to 3 in.	.60	4.50
Tree Surgery	1:3	¼ in.	1 to 4 in.	.65	4.75
Troughs, Water	1:2:3	1 in.	4 to 6 in.	.65	4.75
Walls	1:2:4	1½ in.	4 to 6 in.	.75	5.50
Walls subject to moisture	1:2:3	1½ in.	4 to 6 in.	.65	4.75

TABLE B-8

MAXIMUM SIZE OF AGGREGATE RECOMMENDED FOR VARIOUS TYPES OF CONSTRUCTION

Minimum Dimension of Section (Inches)	Maximum Size of Aggregate*, in Inches, for:			
	Reinforced Walls, Beams, and Columns	Unreinforced Walls	Heavily Reinforced Slabs	Lightly Reinforced or Unreinforced Slabs
2½—5	½—¾	¾	¾—1	¾—1½
6—11	¾—1½	1½	1½	1½—3
12—29	1½—3	3	1½—3	3
30 or more	1½—3	6	1½—3	3—6

*Based on square openings.

TABLE B-10

NUMBER OF SQUARE FEET FROM 1 CUBIC YARD OF CONCRETE

Thickness Inches	No. Sq. Ft.	Thickness Inches	No. Sq. Ft.	Thickness Inches	No. Sq. Ft.	Thickness Inches	No. Sq. Ft.
1	324	4	81	7	46	10	32
1¼	259	4¼	76	7¼	44	10¼	31
1½	216	4½	72	7½	43	10½	31
1¾	185	4¾	68	7¾	42	10¾	30
2	162	5	65	8	40	11	29½
2¼	144	5¼	62	8¼	39	11¼	29
2½	130	5½	59	8½	38	11½	28
2¾	118	5¾	56	8¾	37	11¾	27½
3	108	6	54	9	36	12	27
3¼	100	6¼	52	9¼	35	12¼	26½
3½	93	6½	50	9½	34	12½	26
3¾	86	6¾	48	9¾	33	12¾	25½

FOOTINGS — CUBIC MEASURE PER 100 LINEAR FEET

SIZE	Cubic Feet Concrete Per 100 Lin. Feet	Cubic Yards Concrete Per 100 Lin. Feet
6 x 12	50.00	1.9
6 x 16	65.05	2.4
8 x 12	66.67	2.5
8 x 16	88.89	3.3
8 x 18	100.00	3.7
8 x 20	108.00	4.1
10 x 12	83.33	3.1
10 x 66	111.11	4.1
10 x 18	125.00	4.6
10 x 20	135.00	5.1
12 x 12	100.00	3.7
12 x 16	133.33	4.9
12 x 20	166.67	6.1
12 x 24	200.00	7.4

WALLS — CUBIC MEASURE PER 100 SQUARE FEET

WALL THICKNESS	Per 100 Square Feet Wall Cubic Feet Required	Per 100 Square Feet Wall Cubic Yards Required
4"	33.3	1.24
6"	50.0	1.85
8"	66.7	2.47
10"	83.3	3.09
12"	100.0	3.70

SLABS — CUBIC MEASURE PER 100 SQUARE FEET

THICKNESS	Per 100 Square Feet Slab Cubic Feet of Concrete	Per 100 Square Feet Slab Cubic Yards of Concrete
2"	16.7	.62
3"	25.0	.93
4"	33.3	1.24
5"	41.7	1.55
6"	50.0	1.85

MATERIALS FOR 100 SF OF WALLS, FLOORS, SIDEWALKS, OR ANY SLABS

Concrete Base

Slab Thickness	1:1¾:2¾ Cement Sacks	1:1¾:2¾ Sand Cu. Yd.	1:1¾:2¾ Stone Cu. Yd.	1:2:3 Cement Sacks	1:2:3 Sand Cu. Yd.	1:2:3 Stone Cu. Yd.	1:2:3½ Cement Sacks	1:2:3½ Sand Cu. Yd.	1:2:3½ Stone Cu. Yd.	1:2½:4 Cement Sacks	1:2½:4 Sand Cu. Yd.	1:2½:4 Stone Cu. Yd.	1:3:5 Cement Sacks	1:3:5 Sand Cu. Yd.	1:3:5 Stone Cu. Yd.
2½ in.	5.7	0.36	0.62	5.2	0.40	0.59	4.8	0.37	0.64	4.2	0.40	0.63	3.4	0.39	0.65
3	6.8	0.43	0.74	6.3	0.48	0.71	5.8	0.44	0.76	5.0	0.48	0.75	4.1	0.47	0.78
3½	8.0	0.51	0.86	7.3	0.56	0.83	6.8	0.52	0.90	5.8	0.56	0.88	4.8	0.55	0.92
4	9.1	0.58	0.99	8.4	0.64	0.95	7.7	0.59	1.02	6.6	0.64	1.01	5.5	0.63	1.05
4½	10.3	0.65	1.11	9.4	0.72	1.06	8.7	0.66	1.15	7.5	0.72	1.13	6.1	0.70	1.17
5	11.4	0.73	1.23	10.5	0.80	1.19	9.7	0.74	1.28	8.3	0.80	1.26	6.8	0.79	1.31
5½	12.6	0.80	1.36	11.6	0.88	1.31	10.7	0.82	1.41	9.2	0.88	1.39	7.5	0.87	1.45
6	13.7	0.87	1.48	12.6	0.96	1.42	11.6	0.89	1.54	10.0	0.96	1.52	8.2	0.94	1.57

Wearing or Finish Course

Thickness	1:1½ Cement Sacks	1:1½ Sand Cu. Yd.	1:2 Cement Sacks	1:2 Sand Cu. Yd.	1:1:1 Cement Sacks	1:1:1 Sand Cu. Yd.	1:1:1 Stone Cu. Yd.	1:1:1½ Cement Sacks	1:1:1½ Sand Cu. Yd.	1:1:1½ Stone Cu. Yd.	1:1:2 Cement Sacks	1:1:2 Sand Cu. Yd.	1:1:2 Stone Cu. Yd.
½ in.	2.4	0.13	2.0	0.15	2.1	0.08	0.08	1.8	0.07	0.10	1.6	0.06	0.12
¾	3.6	0.19	2.9	0.22	3.1	0.11	0.11	2.7	0.10	0.15	2.4	0.09	0.18
1	4.8	0.26	3.9	0.29	4.2	0.15	0.15	3.7	0.14	0.20	3.2	0.12	0.24
1¼	6.0	0.33	4.9	0.36	5.2	0.19	0.19	4.6	0.17	0.25	4.1	0.15	0.30
1½	7.2	0.40	5.9	0.43	6.3	0.23	0.23	5.5	0.20	0.30	4.9	0.18	0.36
1¾	8.4	0.46	6.9	0.50	7.3	0.27	0.27	6.4	0.24	0.36	5.7	0.21	0.42
2	9.6	0.53	7.9	0.58	8.3	0.31	0.31	7.3	0.27	0.41	6.5	0.25	0.50

INDEX

Practical References for Builders

National Construction Estimator

Current building costs in dollars and cents for residential, commercial and industrial construction. Prices for every commonly used building material, and the proper labor cost associated with installation of the material. Everything figured out to give you the "in place" cost in seconds. Many time-saving rules of thumb, waste and coverage factors and estimating tables are included. **512 pages, 8½ x 11, $16.00. Revised annually.**

Building Layout

Shows how to use a transit to locate the building on the lot correctly, plan proper grades with minimum excavation, find utility lines and easements, establish correct elevations, lay out accurate foundations and set correct floor heights. Explains planning sewer connections, leveling a foundation out of level, using a story pole and batterboards, working on steep sites, and minimizing excavation costs. **240 pages, 5½ x 8½, $11.75**

Plumber's Exam Preparation Guide

Lists questions like those asked on most plumber's exams. Gives the correct answer to each question, under both the Uniform Plumbing Code and the Standard Plumbing Code — and explains why that answer is correct. Includes questions on system design and layout where a plan drawing is required. Covers plumbing systems (both standard and specialized), gas systems, plumbing isometrics, piping diagrams, and as much plumber's math as the examination requires. Suggests the best ways to prepare for the exam, how and what to study and describes what you can expect on exam day. At the end of the book is a complete sample exam that can predict how you'll do on the real tests. **320 pages, 8½ x 11, $21.00**

Masonry & Concrete Construction

Every aspect of masonry construction is covered, from laying out the building with a transit to constructing chimneys and fireplaces. Explains footing construction, building foundations, laying out a block wall, reinforcing masonry, pouring slabs and sidewalks, coloring concrete, selecting and maintaining forms, using the Jahn Forming System and steel ply forms, and much more. **224 pages, 8½ x 11, $13.50**

Building Cost Manual

Square foot costs for residential, commercial, industrial, and farm buildings. In a few minutes you work up a reliable budget estimate based on the actual materials and design features, area, shape, wall height, number of floors and support requirements. Most important, you include all the important variables that can make any building unique from a cost standpoint. **240 pages, 8½ x 11, $12.00. Revised annually**

Wood-Frame House Construction

From the layout of the outer walls, excavation and formwork, to finish carpentry, and painting, every step of construction is covered in detail with clear illustrations and explanations. Everything the builder needs to know about framing, roofing, siding, insulation and vapor barrier, interior finishing, floor coverings, and stairs. . .complete step by step "how to" information on what goes into building a frame house. **240 pages, 8½ x 11, $11.25. Revised edition**

Construction Superintending

Explains what the "super" should do during every job phase from taking bids to project completion on both heavy and light construction: excavation, foundations, pilings, steelwork, concrete and masonry, carpentry, plumbing, and electrical. Explains scheduling, preparing estimates, record keeping, dealing with subcontractors, and change orders. Includes the charts, forms, and established guidelines every superintendent needs. **240 pages, 8½ x 11, $22.00**

Berger Building Cost File

Labor and material costs needed to estimate major projects: shopping centers and stores, hospitals, educational facilities, office complexes, industrial and institutional buildings, and housing projects. All cost estimates show both the manhours required and the typical crew needed so you can figure the price and schedule the work quickly and easily. **352 pages, 8½ x 11, $30.00**

Reducing Home Building Costs

Explains where significant cost savings are possible and shows how to take advantage of these opportunities. Six chapters show how to reduce foundation, floor, exterior wall, roof, interior and finishing costs. Three chapters show effective ways to avoid problems usually associated with bad weather at the jobsite. Explains how to increase labor productivity. **224 pages, 8½ x 11, $10.25**

Concrete Construction & Estimating

Explains how to estimate the quantity of labor and materials needed, plan the job, erect fiberglass, steel, or prefabricated forms, install shores and scaffolding, handle the concrete into place, set joints, finish and cure the concrete. Every builder who works with concrete should have the reference data, cost estimates, and examples in this practical reference. **571 pages, 5½ x 8½, $17.75**

Construction Estimating Reference Data

Collected in this single volume are the building estimator's 300 most useful estimating reference tables. Labor requirements for nearly every type of construction are included: sitework, concrete work, masonry, steel, carpentry, thermal & moisture protection, doors and windows, finishes, mechanical and electrical. Each section explains in detail the work being estimated and gives the appropriate crew size and equipment needed. Many pages of illustrations, estimating pointers and explanations of the work being estimated are also included. This is an essential reference for every professional construction estimator. **368 pages, 11 x 8½, $18.00**

Contractor's Guide To The Building Code

Explains in plain English exactly what the Uniform Building Code requires and shows how to design and construct residential and light commercial buildings that will pass inspection the first time. Suggests how to work with the inspector to minimize construction costs, what common building short cuts are likely to be cited, and where exceptions are granted. If you've ever had a problem with the code or tried to make sense of the Uniform Code Book, you'll appreciate this essential reference. **312 pages, 5½ x 8½, $16.25**

Rough Carpentry

All rough carpentry is covered in detail: sills, girders, columns, joists, sheathing, ceiling, roof and wall framing, roof trusses, dormers, bay windows, furring and grounds, stairs and insulation. Many of the 24 chapters explain practical code approved methods for saving lumber and time without sacrificing quality. Chapters on columns, headers, rafters, joists and girders show how to use simple engineering principles to select the right lumber dimension for whatever species and grade you are using. **288 pages, 8½ x 11, $14.50**

Excavation and Grading Handbook

The foreman's and superintendent's guide to highway, subdivision and pipeline jobs: how to read plans and survey stake markings, set grade, excavate, compact, pave and lay pipe on nearly any job. Includes hundreds of practical tips, pictures, diagrams and tables that even experienced "pros" should have. **320 pages, 5½ x 8½, $15.25**

Construction Industry Production Manual

Manhour tables developed by professional estimators from hundreds of jobs and all types of construction. Thousands of carefully researched figures, accurate charts and precise tables to give the estimator the information needed to compile an accurate estimate. If you have only one book of labor tables, this is the book to have. **176 pages, 5½ x 8½, $8.00**

Manual of Professional Remodeling

This is the practical manual of professional remodeling written by an experienced and successful remodeling contractor. Shows how to evaluate a job and avoid 30-minute jobs that take all day, what to fix and what to leave alone, and what to watch for in dealing with subcontractors. Includes chapters on calculating space requirements, repairing structural defects, remodeling kitchens, baths, walls and ceilings, doors and windows, floors, roofs, installing fireplaces and chimneys (including built-ins), skylights, and exterior siding. Includes blank forms, checklists, sample contracts, and proposals you can copy and use. **400 pages, 8½ x 11, $18.75**

Stair Builders Handbook

If you know the floor to floor rise, this handbook will give you everything else: the number and dimension of treads and risers, the total run, the correct well hole opening, the angle of incline, the quantity of materials and settings for your framing square for over 3,500 code approved rise and run combinations—several for every 1/8 inch interval from a 3 foot to a 12 foot floor to floor rise. **416 pages, 8½ x 5½, $12.75**

Builder's Office Manual

This manual will show every builder with from 3 to 25 employees the best ways to: organize the office space needed, establish an accurate record-keeping system, create procedures and forms that streamline work, control costs, hire and retain a productive staff, minimize overhead, shop for computer systems, and much more. Explains how to create routine ways of doing all the things that must be done in every construction office in a minimum of time, at lowest cost and with the least supervision possible. **208 pages, 8½ x 11, $13.25**

Plumbers Handbook Revised

This new edition shows what will and what will not pass inspection in drainage, vent, and waste piping, septic tanks, water supply, fire protection, and gas piping systems. All tables, standards, and specifications are completely up-to-date with recent changes in the plumbing code. Covers common layouts for residential work, how to size piping, selecting and hanging fixtures, practical recommendations and trade tips. This book is the approved reference for the plumbing contractors exam in many states. **240 pages, 8½ x 11, $16.75**

Estimating Plumbing Costs

Offers a basic procedure for estimating materials, labor, and direct and indirect costs for residential and commercial plumbing jobs. Explains how to interpret and understand plot plans, design drainage, waste, and vent systems, meet code requirements, and make an accurate take-off for materials and labor. Includes sample cost sheets, manhour production tables, complete illustrations, and all the practical information you need to accurately estimate plumbing costs. **224 pages, 8½ x 11, $17.25**